Concepts and Strategies in Plant Sciences

Series Editor
Chittaranjan Kole, Raja Ramanna Fellow, Government of India
ICAR-National Institute for Plant Biotechnology, Pusa, Delhi, India

This book series highlights the spectacular advances in the concepts, techniques and tools in various areas of plant science. Individual volumes may cover topics like genome editing, phenotyping, molecular pharming, bioremediation, miRNA, fast-track breeding, crop evolution, IPR and farmers' rights, to name just a few. The books will demonstrate how advanced strategies in plant science can be utilized to develop and improve agriculture, ecology and the environment. The series will be of interest to students, scientists and professionals working in the fields of plant genetics, genomics, breeding, biotechnology, and in the related disciplines of plant production, improvement and protection.

Interested in editing a volume? Please contact Prof. Chittaranjan Kole, Series Editor, at ckoleorg@gmail.com

More information about this series at http://www.springer.com/series/16076

Anjanabha Bhattacharya · Vilas Parkhi ·
Bharat Char
Editors

CRISPR/Cas Genome Editing

Strategies And Potential For Crop Improvement

Springer

Editors
Anjanabha Bhattacharya
Mahyco Research Centre
Dawalwadi, Badnapur, Jalna
Maharashtra, India

Vilas Parkhi
Mahyco Research Centre
Dawalwadi, Badnapur, Jalna
Maharashtra, India

Bharat Char
Mahyco Research Centre
Dawalwadi, Badnapur, Jalna
Maharashtra, India

ISSN 2662-3188 ISSN 2662-3196 (electronic)
Concepts and Strategies in Plant Sciences
ISBN 978-3-030-42021-5 ISBN 978-3-030-42022-2 (eBook)
https://doi.org/10.1007/978-3-030-42022-2

This Springer imprint is published by the registered company Springer Nature Switzerland AG
The registered company address is: Gewerbestrasse 11, 6330 Cham, Switzerland

Preface

CRISPR/Cas (**C**lustered **R**egularly **I**nterspersed **S**hort **P**alindromic **R**epeats and **Cas** for CRISPR-associated proteins) is a widely used genome editing (GE) technique among a group which are often collectively referred to as new plant breeding technologies (NPBTs) and can demonstrably have an impact on crop production. CRISPR/Cas can be used as a tool for creating improved crop plants for enhancing the agricultural output and availability of nutrition. Increasing farmer incomes and agricultural productivity to meet the needs of the world's population, which by some estimates, may touch 9.7 billion (bn) by 2020, is a herculean task. Further, climate change, reduced arable land under crop cultivation, water scarcity and reduced labour availability in the agricultural sector will make food production more challenging than ever. GE involves precise alteration of gene sequences without proven unwarranted effects in crops. This book mainly focuses on CRISPR/Cas technology. Among many GE technologies available, CRISPR/Cas, is becoming a tool of choice, due to its simplicity, versatility, scalability and lower cost when compared to other biotechnological tools. By some market estimates, plant breeding and CRISPR/Cas plant market may reach to US $14.55 bn by 2023 thus emphasising the vast potential of this technology in years to come. Random mutagenesis based genome editing through physical irradiation and chemically induced, has been accepted all over the globe but is time-consuming and therefore, may not fulfil emerging market demand. It is, therefore, expected that CRISPR/Cas genome editing will be perceived more positively and therefore, this book is timely. Though traditional breeding has played an important role in bringing green revolution; yields have plateaued to a large extent in many key agricultural crops. A lot in the past depended on the availability of naturally occurring useful genes, some of which might have come from wild relatives through rounds of positive selection. At the same time, a few naturally available alleles might have also cropped up through spontaneous mutation over time.

Given the current importance of genome editing, particularly CRISPR/Cas, it is hoped that the target audience will gain from this book and their combined effort will further facilitate the field of CRISPR/Cas and consumer acceptance of GE products in agriculture. It is imperative that the public must be informed, illustrating, the similarities and differences between genome-edited crops from conventionally bred crops, to increase acceptance. Public outreach is also one of the objectives of this book.

We hope a bit of illustration on the contents will enhance the book's appeal to the audience. Firstly, the major objective of this volume is to introduce the topic to genome editing, editing types and focuses mostly on the popular CRISPR/Cas technique. The CRISPR/Cas field is vast and almost daily new findings are reported. We have tried our best to include as much information as possible to keep the contents up-to-date. However, it is far from complete by all means. The book begins with a journey on CRISPR/Cas through Chap. 1, which is written by the editors themselves, focuses on genome editing, glimpses on the history of genome editing, classification, basic experimental application followed by genome editing in agricultural crops and challenges it faces. Some of the topics have been described in detail in subsequent chapters. Chapter 2 provides detailed insight on the tools used for designing CRISPR/Cas experiments, bioinformatics aspects, which will be very useful for students, faculties and researchers alike. Chapter 3 provides detailed deliberation on food security by emphasising the role of genes attributing to yield, quality, tolerance to abiotic and biotic stress, and accelerated domestication of crops. Then, Chapter 4 takes us through genetic engineering of floricultural crops. Usually, gene editing in floriculture crops takes a back seat compared to other field and vegetable crops. Therefore, it is our endeavour to include the topic specifically in this book. Chapter 5 discusses applications of CRISPR/Cas to combat multiple abiotic stresses in a crop cycle by modifying key genes. In Chap. 6, a description of use of CRISPR/Cas in vegetable crop improvement is elaborated. We hope this will be very useful to the crop improvement community and draw attention about the usefulness of the technology among vegetable breeders. Further, Chap. 7 focuses on CRISPR/Cas application in climate start agriculture. This is particularly important as climate change is seriously threatening crop production, putting human lives at the risk of impending starvation and malnutrition. A brief outline of the classical crop improvement techniques, proof of concept studies and traits influenced by climatic regimes are discussed. Chapter 8 deliberates on the translational research aspect of CRISPR/Cas in plant genomic research using freely available tools. The application of GE in plant breeding is also discussed for achieving robust plant architecture, flowering and fruit maturation. While, Chap. 9, walks us through the regulatory framework and policy decisions which may finally decide the success of CRISPR/Cas9-GE and positively affect consumer sentiments. Importantly, the chapter deliberates on public understanding of the technology and regulatory acceptance perspective from several nations. Therefore, improving the public image is the key for genome-edited crops to succeed. At the same time, it is imperative that timely approval of biotechnology enabled products and robust regulatory mechanisms must exist so as to facilitate the development of a new generation of gene-edited crops by different stakeholders including start-ups working in this arena. The unwarranted steep regulatory burden puts the additional economic burden to the final product, thus preventing public players from bringing genome-edited to market in a time-bound manner and thwarts private players from investing in new technologies. Chapter 10 focuses on field and cereal crop development. Finally, Chap. 11 walks us through the intellectual property landscape involved with the usage of this technology.

Our readers, at times, may observe a bit of overlap in some topics deliberated in individual chapters. We felt, CRISPR/Cas technique cannot be seen in isolation, and some overlap is completely unavoidable in the context of the topic discussed. This repetitive approach will also help to enforce our understanding of concepts and topic from a different view-point, narrated by individual author and their team for the particular chapter. Finally, we conclude by advocating CRISPR/Cas and associated NPBTs could bring hope for the second wave of the green revolution. This combined with the availability of next-generation sequencing data can help us to create designer crops for the future and defray the ill effects of climate change on global food production to a large extent.

Thus, we hope, this book will provide readers a holistic view of CRISPR/Cas technology and its application in crop plants. We thank Mahyco (Maharashtra Hybrid Seeds Company, Jalna, India), for allowing us to take up this book project and look forward to all suggestions from our readers in making future improvements to the edition. Sincere, special thanks to Dr. Chittaranjan Kole, Series editor for his ever encouragement and team Springer who whole-heartedly accepted our proposal to edit a book project on this important topic of CRISPR/Cas. We sincerely thank all contributors spanning several countries and continents for making this project a success. Thus, for all our combined effort, this book project is able to see the light of the day. We would finally like to dedicate this book to our beloved founder Chairman, Late Dr. B. R. Barwale, World Food Prize Laureate, who was always a strong proponent for use of new plant breeding technologies in modern Agriculture for ultimately improving the livelihood of millions of farmers in India and worldwide.

We wish everyone a happy read!

Jalna, India

Anjanabha Bhattacharya
Vilas Parkhi
Bharat Char

Aims and Scope

CRISPR/Cas genome editing (GE) which is referred as new plant breeding technologies (NBTs), can be used as an useful tool for creating improved crop plants. GE involves precise alteration of gene sequences without unwarranted effects in crops. Though traditional breeding has played an important role in bringing green revolution, yields have plateaued to a large extent in many key agricultural crops. NBTs bring hope for a second wave of green revolution. This is particularly important as climate change is seriously threatening crop production, putting human lives at the risk of impending starvation and malnutrition. Therefore, this book aims to introduce readers to GE, various concepts and tools used in GE, and discuss its proven use in modifying agricultural traits in key crops. We will discuss IP (intellectual property) scenario, which is miraged with over-lapping patents and conflicts. Application of CRISPR technology in translational research is also discussed. Lastly, regulatory framework and policy decisions are discussed in light of commercialization of gene edited crops. Thus, this book will provide readers a holistic view of CRISPR/Cas technology and its application in crop plants.

Jalna, India

Anjanabha Bhattacharya
anjanabha.bhattacharya@mahyco.com
Vilas Parkhi
vilas.parkhi@mahyco.com
Bharat Char
bharat.char@mahyco.com

Contents

1 **Introduction to Genome Editing Techniques: Implications in Modern Agriculture** 1
Anjanabha Bhattacharya, Vilas Parkhi, and Bharat Char

2 **Application of Bioinformatics Tools in CRISPR/Cas** 31
Shalu Choudhary, Abhijit Ubale, Jayendra Padiya, and Venugopal Mikkilineni

3 **CRISPR and Food Security: Applications in Cereal Crops** 53
Mayank Rai, P. Magudeeswari, and Wricha Tyagi

4 **Improvement of Floriculture Crops Using Genetic Modification and Genome Editing Techniques** 69
Ayan Sadhukhan and Heqiang Huo

5 **Enhancing Abiotic Stress Tolerance in Plants Through Genome Editing** 91
Deepa Jaganathan, Rohit Kambale, Hifzur Rahman, Devanand Pachanoor Subbian, and Raveendran Muthurajan

6 **Vegetable Crop Improvement Using CRISPR/Cas9** 119
Francisco F. Nunez de Caceres Gonzalez and Daniela De la Mora Franco

7 **Use of CRISPR in Climate Smart/Resilient Agriculture** 131
Vinod Kumar, Sabah AlMomin, Muhammad Hafizur Rahman, and Anisha Shajan

8 **Translational Research Using CRISPR/Cas** 165
Anshika Tyagi, Sandhya Sharma, Sanskriti Vats, Sajad Ali, Sandeep Kumar, Naveed Gulzar, and Ruspesh Deshmukh

9 **Regulatory Framework and Policy Decisions for Genome-Edited Crops** 193
Anirudh Kumar, Rakesh Kumar, Nitesh Singh, and Aadil Mansoori

10 Field Crop Improvement Using CRISPR/Cas9 203
Elangovan Mani

11 Patent Landscape of CRISPR/Cas 213
Paramita Ghosh

Contributors

Sajad Ali Center of Research for Development, University of Kashmir, Srinagar, India

Sabah AlMomin Biotechnology Program, Environment and Life Sciences Research Center, Kuwait Institute for Scientific Research, Safat, Kuwait

Anjanabha Bhattacharya Mahyco Research Centre, Maharashtra Hybrid Seed Company Private Limited, Dawalwadi, Jalna, Maharashtra, India

Bharat Char Mahyco Research Centre, Maharashtra Hybrid Seed Company Private Limited, Dawalwadi, Jalna, Maharashtra, India

Shalu Choudhary Mahyco Private Limited, Dawalwadi, Jalna, Maharashtra, India

Daniela De la Mora Franco CINVESTAV, Irapuato, Mexico

Ruspesh Deshmukh National Agri-food Biotechnology Institute, Mohali, India

Paramita Ghosh Mahyco Research Centre, Dawalwadi, Jalna, Maharashtra, India

Naveed Gulzar Center of Research for Development, University of Kashmir, Srinagar, India

Heqiang Huo Mid Florida Research and Education Centre, University of Florida, Apopka, FL, USA

Deepa Jaganathan Centre for Plant Molecular Biology and Biotechnology, Tamil Nadu Agricultural University, Coimbatore, India

Rohit Kambale Centre for Plant Molecular Biology and Biotechnology, Tamil Nadu Agricultural University, Coimbatore, India

Anirudh Kumar Department of Botany, Indira Gandhi National Tribal University (IGNTU), Amarkantak, Madhya Pradesh, India

Rakesh Kumar Department of Life Science, Central University of Karnataka, Kalaburagi, Karnataka, India

Sandeep Kumar Xcelris Labs Ltd., Ahmedabad, India

Vinod Kumar Biotechnology Program, Environment and Life Sciences Research Center, Kuwait Institute for Scientific Research, Safat, Kuwait

P. Magudeeswari School of Crop Improvement, College of Post Graduate Studies in Agricultural Sciences, Central Agricultural University (Imphal), Umiam, Meghalaya, India

Elangovan Mani Advanta Seeds, UPL Ltd., Hyderabad, India

Aadil Mansoori Department of Botany, Indira Gandhi National Tribal University (IGNTU), Amarkantak, Madhya Pradesh, India

Venugopal Mikkilineni Mahyco Private Limited, Dawalwadi, Jalna, Maharashtra, India

Raveendran Muthurajan Centre for Plant Molecular Biology and Biotechnology, Tamil Nadu Agricultural University, Coimbatore, India

Francisco F. Nunez de Caceres Gonzalez Bayer Vegetable Seeds, Almeria, Spain

Jayendra Padiya Mahyco Private Limited, Dawalwadi, Jalna, Maharashtra, India

Vilas Parkhi Mahyco Research Centre, Maharashtra Hybrid Seed Company Private Limited, Dawalwadi, Jalna, Maharashtra, India

Hifzur Rahman Centre for Plant Molecular Biology and Biotechnology, Tamil Nadu Agricultural University, Coimbatore, India

Muhammad Hafizur Rahman Biotechnology Program, Environment and Life Sciences Research Center, Kuwait Institute for Scientific Research, Safat, Kuwait

Mayank Rai School of Crop Improvement, College of Post Graduate Studies in Agricultural Sciences, Central Agricultural University (Imphal), Umiam, Meghalaya, India

Ayan Sadhukhan Mid Florida Research and Education Centre, University of Florida, Apopka, FL, USA

Anisha Shajan Biotechnology Program, Environment and Life Sciences Research Center, Kuwait Institute for Scientific Research, Safat, Kuwait

Sandhya Sharma ICAR-National Institute for Plant Biotechnology, New Delhi, India

Nitesh Singh Department of Botany, Indira Gandhi National Tribal University (IGNTU), Amarkantak, Madhya Pradesh, India

Devanand Pachanoor Subbian Centre for Plant Molecular Biology and Biotechnology, Tamil Nadu Agricultural University, Coimbatore, India

Anshika Tyagi ICAR-National Institute for Plant Biotechnology, New Delhi, India

Wricha Tyagi School of Crop Improvement, College of Post Graduate Studies in Agricultural Sciences, Central Agricultural University (Imphal), Umiam, Meghalaya, India

Abhijit Ubale Mahyco Private Limited, Dawalwadi, Jalna, Maharashtra, India

Sanskriti Vats National Agri-food Biotechnology Institute, Mohali, India

Abbreviations

AAAS	American Association for the Advancement of Science
AAV	Adenovirus
ABA	Abscisic acid
ABE	Adenine base editor
ACC	1-aminocyclopropane-1-carboxylic acid
ACS	ACC synthase
AFLP	Amplified fragment length polymorphism
ALS	Acetolactate synthase
AOP	Apomictic offspring producer
BADH2	Betaine aldehyde dehydrogenase
Bbm	Baby boom
bp	Base pair
Bt	Bacillus thuringiensis
CAD	Cinnamyl alcohol dehydrogenase
CaMV	Cauliflower mosaic virus
Cas	CRISPR associated protein
CBE	Cytidine base editor
CCD4	Carotenoid cleavage dioxygenase 4
CCD7	Carotenoid cleavage dioxygenase 7
CHLI1	Magnesium-chelatase subunit I
CKX2	Cytokinin oxidase or dehydrogenase
ClCuV	Cotton leaf curl virus
CRISPR	Clustered Regularly Interspersed Short Palindromic Repeat
CRISPRi/a	CRISPR inhibition or activation
CRISTO	Carotenoid isomerise
CSIT	Critical Sterility Inducing Temperature
CTR1	CONSTITUTIVE TRIPLE RESPONSE 1
CVYV	Cucumber vein yellowing virus
dCas9	Deactivated/dead Cas9
DEP1	Dense and Erect Panicle
DMR	Downy mildew resistance
DNA	Deoxyribonucleic acid

DREB2	Dehydration responsive element-binding protein 2
DSB	Double-strand break
eIF4G	Translation Initiation Factor 4 Gamma gene
EIN2	ETHYLENE INSENSITIVE2
ENGase	Endo-N-acetylb-D-glucosaminidase
EPO	European Patent Office
EPSPS	5-enolpyruvylshikimate-3-phosphate
ERFs	ETHYLENE RESPONSEFACTORs
EU	European Union
FAD	Fatty acid desaturase
FAO	Food and Agriculture Organisation
GBSS	Granule-bound starch synthase
GE	Genome Editing
GEEN	Genome Editing with Engineered Nucleases
Ghd7	Grains Height Date 7
GM	Genetic modification
GMO	Genetically Modified Organism
Gn1a	Grain Number 1a
GS3	Grain size 3
Gus	ß-glucuronidase
GVR	Geminivirus Replicon
GW	Grain Width
GW2	Grain weight 2
HAL3	Halo tolerance protein
HDR	Homology-Directed Repair
HD-Zip	Homeodomain-leucine zipper protein
HR	Homologous recombination
HSP	Heat shock protein
Htd1	High tillering date 1
IAEA	International Atomic Energy Agency
IPA1	Ideal Plant Architecture 1
IPK	Inositol phosphate kinase
ISSR	Inter simple sequence repeats
KO	Knock out
LOB1	LATERAL ORGAN BOUNDARIES 1
MAPK3	Mitogen active protein kinase 3
MAPKs	Mitogen-activated protein kinases
MAS	Marker-assisted selection
miRNAs	MicroRNAs
MIT	Massachusetts Institute of Technology
MLO	Mildew Resistance Locus
mRNA	Messenger RNA
MS	Murashige and Skoog
NAA	Napthalene acetic acid
NBTs	New breeding techniques

NCED 1	Phaiustankervilliae9-cis-epoxycarotenoid dioxygenase NCED1
NGS	Next-Generation Sequencing
NHEJ	Non-homologous End Joining
NTWG	New Techniques Working Group
ODM	Oligonucleotide Directed Mutagenesis
OsANN3	rice annexin3 gene
OsDERF1	Drought-responsive ERF gene
OsPMS	Protein mismatch repair
OSR	Organ Size Regulation
P5CS	Pyrroline-5-Carboxylate Synthase
PAM	Protospacer-adjacent motif
PB	Precision breeding
PDR6	Pleiotropic drug resistance 6
PDS	Phytoene desaturase
PEG	polyethylene glycol
PPO	Polyphenol oxidase
PPRs	Pentatricopeptide repeat proteins
PRSV-W	Papaya ringspot mosaic virus type-W
PTAB	Patent Trial and Appeal Board
QTL	Quantative trait loci
RAPD	Random amplified polymorphic DNA
RK2	SNF 1-RELATED PROTEIN KINASE 2
RNAi	interference RNA
RNPs	Ribonucleoproteins
ROS	Reactive oxygen species
RR22	*Oryza sativa* Response Regulator 22
RTBV	Rice tungro bacilliform virus
RTSV	Rice tungro spherical virus
SAGs	Senescence-Associated Genes
SAM	S-adenosyl-L-methionine
SBA	Swedish Board of Agriculture
SBEs	Starch Branching Enzyme
SCR	SCARECROW
SDN	Site-Directed Nuclease
SDNs	Site-directed nucleases
sgRNAs	single guide RNAs
SHR	SHORT-ROOT
SKC1	Shoot K^+ Concentration 1
SlAGL6	Slagamous like 6
SLAGO7	SLARGONAUTE
SNAC2	Stress-responsive NAC gene
SpCas9	*Streptococcus pyogenes*
SPL16	Squamosa Promoter Binding Protein like 16
SSR	Simple sequence repeat
TALEN	Transcriptional Activator-Like Effector Nuclease

TB1	Teosinte Branched 1
TGMS	Thermo-sensitive genetic male sterility
TGW6	Thousand-grain weight 6
tracrRNAs	trans-activating crRNAs
TSS	Transcription Start Site
TT4	TRANSPARENT TESTA4
TuMV	Turnip mosaic virus
UCB	University of California, Berkley
USPTO	United State Patent and Trademark Office
VIGS	Virus-induced gene silencing
Wus 2	Wuschel 2
ZFN	Zinc Finger Nuclease
ZYMV	Zucchini yellow mosaic virus

Chapter 1
Introduction to Genome Editing Techniques: Implications in Modern Agriculture

Anjanabha Bhattacharya, Vilas Parkhi, and Bharat Char

Abstract Climate change accompanied by global warming is happening at an ever alarming pace and is here to stay. Therefore, food production challenges to feed our ever-burgeoning population, which is expected to rise to 9.7 billion by 2050, will remain a steep challenge. Conventional breeding which has created a wave for the first green revolution may not come to our rescue this time entirely alone, in meeting the herculean goal. Therefore, for the second green revolution, agricultural biotechnology-based tools including precision breeding have to play a critical role. For genome editing (GE) induced precision breeding to succeed strong political, social and cultural support is needed. Changing public perception by providing complete information about genome editing will boost consumer confidence and encourage the consumption of GE food crops. Among many GE technologies available, CRISPR/Cas (**C**lustered **R**egularly **I**nterspersed **S**hort **P**alindromic **R**epeats and **Cas** for CRISPR associated proteins), is becoming a tool of choice, due to its simplicity, versatility, scalability and lower cost when compared to other biotechnological tools. Improved versions of CRISPR/Cas-GE, such as base editing are coming to the centre stage and many more technological breakthroughs are expected in the near future. Random mutagenesis based genome editing through physical irradiation and chemically induced, has been accepted all over the globe and it is, therefore, expected that CRISPR/Cas genome editing will be perceived positively. This combined with the availability of next-generation sequencing data can help us to create designer crops for the future and defray the ill effects of climate change on global food production to a large extent.

Keywords Genome editing (GE) · CRISPR/Cas9 · Base editing · Agricultural crops

A. Bhattacharya (✉) · V. Parkhi · B. Char (✉)
Mahyco Research Centre, Maharashtra Hybrid Seed Company Private Limited, Aurangabad-Jalna Road, Dawalwadi, Jalna 431203, Maharashtra, India
e-mail: anjanabha.bhattacharya@mahyco.com

B. Char
e-mail: bharat.char@mahyco.com

A. Bhattacharya et al. (eds.), *CRISPR/Cas Genome Editing*, Concepts and Strategies in Plant Sciences, https://doi.org/10.1007/978-3-030-42022-2_1

1.1 Introduction

Conventional crop improvement is a time-consuming process, which is an interplay of random permutation and combinations to ultimately achieve cultivars with improved yield, biotic, abiotic resistance and quality traits among other economical traits in an era of rapid climate change. However, domestication of crop plants for the last hundred years has resulted in selecting germplasms with narrow genetic base and domestication for predominantly yield traits (Ross-Ibarra et al. 2007; Zhang et al. 2017a, b, Vaughan et al. 2017; Smykal et al. 2018). This has resulted in stagnating yield curve barrier which is difficult to break by conventional breeding techniques alone. This has led to heightening focus on other biotechnology-based interventions including precision breeding (PB) or new breeding techniques (NBTs) (Aglawe et al. 2008). NBTs are broadly collection of techniques which can be efficiently used for crop improvement. Broadly, these include all types of genome editing techniques, genomic assisted tools, gene methylation, *Agro*-infiltration, reverse breeding, oligonucleotide-directed mutagenesis (ODM), single-stranded nucleotide directed editing, SSDN and other molecular biology tools (Schaart et al. 2016; Yoshmi et al. 2016; Mohanta et al. 2017). Oerke (2006) predicted global loss in crop productivity is around 34%, though by some recent estimates is close to 40% (website 1). This might make a lofty goal to feed the World's ever-burgeoning population, which is expected to rise to 9.7 billion by 2050 (website 2). Therefore, it is expected that trait developed through NPBTs can deliver and meet consumer demand. GE (genome-edited) crops can result in the development of traits which, otherwise could be achieved, in only selected cases, by conventional breeding, over several years. While for other traits, conventional breeding may not suffice. Thus, GE helps to meet present demand in a much shorter time period (Duensing et al. 2018). To make our point in favour of precise GE, random mutagenesis/random genome editing in agricultural crops has resulted in the release of more than 2500 cultivars in 175 plant species, which shows that random mutagenesis-based genome editing through irradiation and chemically induced, has been accepted all over the globe (website 3—FAO/IAEA FAO/IAEA 2014; Songstad et al. 2017) and it is expected that present approach of using CRISPR/Cas9 for genome editing will be perceived in a more positive way because of its precise nature. As pointed out in a paper by Wolter and Puchta (2017) that if the plant has one to a few nucleotide change and is transgene-free, it cannot be differentiated experimentally from the one which originated by other means of mutagenesis. For precision breeding to succeed strong political, social and cultural support is needed. Changing public perception in European and other non-GM markets by providing complete information about genome-edited products and cultivating confidence will ultimately decide the future of these new generations of biotechnology-assisted products for crop improvement. Presently, there is a broad spectrum of technologies, that is conventional breeding to GM and now GE, in use for improving agriculture, each of these has already drawn heavy investment from stakeholders. Therefore, not only scientific basis but political, social, economic consideration and institutional will are needed to capture the usefulness of such technologies. At present, there is a contrasting difference in view-point between European and American consumers in relation to consumption of biotech derived crops. While the American consumer has embraced GM for decades, EU nations have adopted a cautious approach (please see, Lassoued et al. 2018 for detailed results of the social survey on agriculture biotechnology crops). Therefore, improving the public image is the key for genome-edited crops to succeed.

In this chapter, the term CRISPR/Cas includes all variants of Cas including Cas 1, Cas 3, Cas 4, Cas9, Cpf1, Csm, Cas 13. Further, our focus in the present book chapter will be primarily on CRISPR/Cas technology among other GE techniques. Therefore, detailed deliberation will be done involving, CRISPR/Cas technology, mostly due to its wider acceptability, ease of design, scalability, simplicity, speed and overall low cost. While a brief overview will cover other genome editing options available to us, to develop better agricultural crop traits. There are four types of edits which are possible by employing GE, namely, knock out or deactivate a gene, activate or overexpress, repress and precisely alter gene including targeted chromosome rearrangements (Choi and Meyerson 2014). The outcome of GE could result in homozygous, heterozygous, biallelic and mosaic mutations. Broadly, genome editing is divided into four classes according to their chronological order of discovery. Initially, (1) Meganucleases (MNs) or customized homing nucleases were used in gene editing followed by (2) Zinc finger nucleases (ZNFs), (3) TALEN and (4) CRISPR/Cas (Fig. 1.1a–d). While MNs, ZNFs and TALENs are protein-based, CRISPR/Cas is an RNA–protein

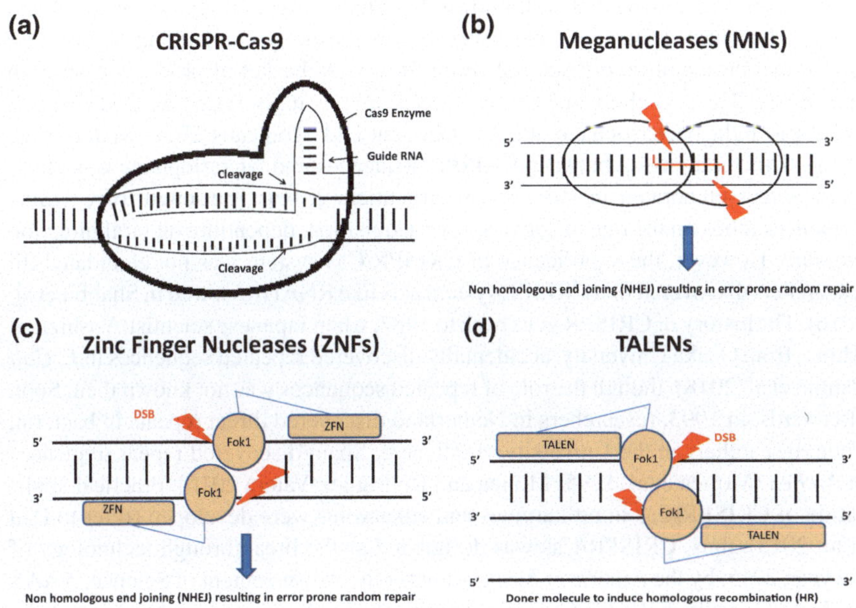

Fig. 1.1 Four class of GE tools

hybrid system. We will describe each of these technologies one by one starting with CRISPR/Cas. Essentially, GE mainly focuses on precise genetic manipulation of a target sequence, but many other application including transcriptional regulation, genomic screens, imaging and diagnostics is also possible.

1.2 History and Origin of CRISPR/Cas

CRISPR stands for **C**lustered **R**egularly **Interspersed** **S**hort **P**alindromic **R**epeats and **Cas** for CRISPR associated proteins. This is basically, an acquired immune system of bacteria and archaea against invading viruses (Barrangou and Marraffini 2014). Bacteria are in constant war with invading viral particles. Thus, in order for bacteria to survive a plethora of viral attack, bacteria developed an innate immune system, termed as CRISPR/Cas, a two-component system. The system degrades foreign DNA that re-enters the cell. Once a virus attacks bacteria, built-in CRISPR system copies segment of the repeats of the viral genome which are few base pairs long of extrachromosomal origin, separated by non-coding sequences known as spacer sequences (Bolotin et al. 2005) adjacent to PAM (Protospacer Activating Motifs) (Karginov and Hannon 2010). When the second round of attack takes place, the CRISPR/Cas system is activated and degrades the viral genome. In short, CRISPR/Cas is remembered and destroys the system. The Cas library not only maintains an array of records of invading viral signature but also helps to destroy upon re-infection. Previously, it is well known that phage infection is a serious problem in dairy business (website 4) and it became increasingly challenging to maintain an original pure culture of bacterial strain for a wide basket of products offered to consumers. Thus, yoghurt makers have been unknowingly relied on CRISPR/Cas system to fight bacteriophage attacks (Garneau and Moineaus 2011; Marco et al. 2012). This relationship between CRISPR sequences and bacteriophage was elucidated and is elaborated in Hille and Charpentier (2016). The length of CRISPR sequences relies on the rate of active spacer acquisition, depending on viral infection pressure. However, the significance of CRISPR/Cas system was not elucidated till researchers discovered that CRISPR system acts like RNAi (reviewed in Shabbir et al. 2016). The history of CRISPR goes back to 1987, when Japanese scientist, Yoshizumi Ishino, from Osaka University, accidentally discovered repeated sequences in *E. Coli* (Ishino et al. 2018), though the role of repeated sequences was not known then. Soon afterwards, in 1993, researchers in Netherland discovered direct repeats in bacteria. Then, researchers in the University of Alicente, Spain discovered repeat sequences in *Archea* (Mojica et al. 1995; Mojica and Rodriguez-Valera 2016). Practical application of CRISPR/Cas in prokaryotes and eukaryotes were developed (refer to Hsu et al. 2015), thus, CRISPR/Cas was designated as the breakthrough technology of the year, 2015, by the American Association for the Advancement of Science, AAAS (website 5). Truly, CRISPR/Cas is a disruptive technology involved in rewriting the genetic code. Some believe that CRISPR/Cas technology is superior to RNAi (interference RNA, a gene silencing technique), as it can create knock out (KO) while RNAi cannot fully suppress gene activity under a varying array of environmental conditions (Boettcher and McManus 2015; Unniyampurath et al. 2016; Arora and Narula 2017). Here, it must be mentioned that while bacteria have developed CRISPR/Cas system to defend themselves against phage/ virus, the phage itself has counter developed anti-CRISPR/Cas strategy. This involves phage mutating CRISPR library thereby evading destruction or have developed protein which prevents Cas binding. Anti-CRISPRs are also common in mobile genetic elements. The anti-CRIPR proteins are small enough to pass through the nuclear pores (Pawluk et al. 2016). However, anti-CRISPR strategies are few and far in between and CRISPR/Cas system is likely to dominate the field of genome editing (see Pawluk et al. 2018).

1.3 Mode of Action of CRISPR/Cas System

CRISPR/Cas is an RNA-guided nuclease which edits the DNA sequence. Generally, U3/U6 (RNA pol III promoter) is used to drive SgRNA, while CaMV35S and ubiquitin promoter is used to drive Cas9. Both gRNA and Cas must be expressed in the cell at the same time for the editing to take place. The most commonly studied system is CRISPR/Cas9 from *Streptococcus pyrogens* (type-II) and Cpf1 (type V). It must be noted that Cas9 is better than Cas type-I, which employ multiple subunits rather than a single one. The **CRISPR/Cas** system comprises of monomeric DNA non-specific endonuclease (Cas 9) and customizable single guide RNA (synthetic guide RNA). The guide RNA is a two-component system and contains CRISPR RNA (crRNA) and transactivating (trRNA, Deltcheva et al. 2011), which direct Cas9 protein to the target sequence and results in site-specific cut, generally in the vicinity of 3–4 bp upstream of PAM (protospacer adjacent motif) site. The PAM is necessary for Cas9 binding to the DNA sequence, adjacent to the sequence sgRNA represent (Jinek et al. 2012). For the formation of mature guides from pre-cr RNA to happen, processing and transcription take place from the DNA sequence. For clarity, we will use the term g or Sg-RNA for the synthetic fusion of cr-RNA and trancr-RNA, that is target sequence with scaffold sequence for ease of understanding. This guide RNA facilitates editing and is easy to design. Therefore, guides are essential as well as the star feature of the CRISPR/Cas system. Editing takes place by the formation of RNA–DNA complex, followed a repair mechanism that uses template/donor DNA, causing DNA cleavage by Cas9 complex (Huai et al. 2017; Sundaresan et al. 2017) which is subsequently repaired by endogenous repair mechanism. In order to increase the efficiency of editing, interaction of non-target strand, positively charged groove of the Cas complex and gRNA with target strand should be optimized (Guha et al. 2017). This can be achieved by weakening the bond in the Cas complex, thus increasing specificity. The DNA repair is of three types, that is error prone repair of DNA occurs by NHEJ (nonhomologous end joining, Schiml et al. 2014) and precise edits which, happens by DSB (double-stranded break in both leading and lagging strand, Anders et al. 2014) induced by homologous directed repair (HDR). CRISPR/Cas induced HR (homologous recombination) may involve biolistics, as this helps to increase copy number and facilitate HR process. The third one is micro-homology mediated end-joining (MMEJ) is a repair process involving very short homology regions (5–25 bp) flanking DSB (for details, see, Seol et al. 2018). The system is limited by the availability of the PAM region in the gene of interest. HDR mediated insertions have been achieved by *Agrobacterium*-mediated transformation. Cas9 has three domains, the larger one being REC connected to the smaller one (RuvC) and the third HNH-nuclease domain, which results in a positively charged groove and binds to PAM sequence. An added advantage of CRISPR/Cas9 is that it can be used to simultaneously edit several genes (Ma et al. 2015; Li et al. 2018a, b, c; Sekine et al. 2018). CRISPR/Cas system themselves do not have any special preference for coding and non-coding sequences (Canver et al. 2017). Another Cas protein, Cas 12a (Cpf1) is derived from *Provoltella, Francisella*-1 (Zetsche et al. 2015) with the recognition, T rich, TTTN PAM site. Cpf1-gRNA is much smaller, 42nt long compared with Cas9 which is 92 bp long and Cpf1 has high multiplexing capability compared to Cas 9 (Li et al. 2018a, b, c). Cpf1 locks HNH endonuclease domain and needs one RNA instead of both cr- and transcr-RNA for editing. Cpf1 is much smaller than Cas9, is preferably used in HDR and gene insertion or replacement. Cpf1 induces sticky overhangs of 5 bp, 18–23 bp away from PAM site whereas Cas9 induces blunt cut, 3 bp upstream of PAM site (refer to Deng et al. 2018). Cpf1 is better at editing non-dividing cells. Since, Cpf1 results in staggered cut way from PAM site, it is retained for editing, sometimes referred as second chance editing, facilitating HDR repair (for details refer Zetche et al. 2015). However, it should be ensured that once HDR is complete, the homologous sequence or repair template must have an inactive PAM site to prevent the possibility of subsequent editing. Cpf1 can facilitate self-cleavage activity involving multiple genes and therefore is better in editing genes than Cas9 (Li et al. 2018a, b, c). Cpf1 lacks HNH therefore, conversion to nickase is not possible. Three other orthologs of Cpf1 are available from *Francisella novicida* (FnCas12a), *Lachnospiraceae bacterium* ND 2006, *Acidaminococcus* sp BV3L6 (Verwaal et al. 2017). Dead Cas9, can be used to develop inactive versions of the target gene (CRISPRi, Dominguez et al. 2016) which can be used in epigenomic editing (Qi et al. 2013; Larson et al. 2013). The potential application of dCas9 may include, modulating the expression pattern of gene, using a modified version of Cas9. Another point to note is that dead Cas versions used for activation or repression do not cause off-target editing problems. The development of Cas9nickase results in cleavage of only one strand of DNA and thus lowering the frequency of off-target editing (Sander and Joung 2014). Nickase is appropriate in directing sequence tags and mutations employing GE studies. Enhanced eSpCas9 and eSpCas9-HF1 are available which reduces Cas activity and at the same time increasing sensitivity to specific edits in the target gene (Slaymaker et al. 2016). Engineered versions of Cas are now available for precision editing. Further, the availability of multiple Cas9 orthologs facilitates individual Cas9s to be used together in one experiment (Tycko et al. 2017). Alternatively, there is another approach to deliver CRISPR/Cas into the cell, which is based on the protein-RNA complex, thereby bypassing transcription and translation associated with DNA editing. The alternative technique is known as

DNA free editing that is RNP techniques (Yin et al. 2017a, b). Cas can be delivered as protein and sgRNA as m-RNA known as RNP (ribonucleoprotein). In RNP experiments, care must be undertaken to express both individual components at the same time for the success of the experiment. The advantage of using RNP includes the ability to cleave the target without further modification (Kimberland et al. 2018). The complex also degrades very rapidly leaving no residue for any off-target editing. Analysis of mutations is becoming easier day by day. Many of the bio-informatic tools such as DSDECODE and TIDE can classify mutation types from the original PCR product, while NGS can do the same with a precision of 0.01%. Tools like Hi-TOM enables tracking of precise mutations and is particularly helpful to researchers who are unfamiliar with bioinformatics.

1.4 Classification of CRISPR/Cas

There is broadly, Class 1 CRISPR/Cas, which consists of multiple Cas proteins, again divided into three more sub-classes I, III, IV. Class 2, which consists of single Cas protein and is divided into type II (Cas9), IIA (Csn2), IIB (Cas4), V (Cpf1, C2c1, C2C3), VI (Cas 13a–c). Further, there are six types and 19 sub-types of CRISPR/Cas system (Makarova and Koonin 2015). In case of multiple CRISPR/Cas proteins, it seems Cas share a conserved domain whereas effector genes have evolved independently, which now we know as guide RNAs. The conserved domain remains interspersed across Cas proteins (for example in Cas 9, Cas 12, Cas 13), please refer to Makarova et al. 2018 for detailed overview. Further, six new programmable class II DNA nucleases (Cpf1 renamed as Cas 12a, C2C1-Cas 12b, C2 C3, Cas Y-12d, Cas X-12c and two RNA targeting systems namely C2C2-Cas 13a (Class V), C2C6 (class VI) (Xie and Yang, 2013). Abudayyeh et al. (2016) reported RNA editing by Cas13a (C2C2) derived from *Leptotrichia shahii*. The RNA editor requires protospacer flanking sites (PFS) instead of PAM to induce breaks (Murovec et al. 2017). Thus, Cas 13a can become an alternate tool for RNAi. RNA binding Cas9 (RCas9) binds and cuts RNA molecule to induce post-transcriptional gene silencing (for more info on Cas 13, please refer to Zhang lab, https://zlab.bio/cas13/). The RCas9 is an addition over dead Cas9 and Cas9-fusion proteins which have a different application of the CRISPR-Cas. The technique can be used for enriching transcripts and used in chromosomal imaging.

1.5 Points to Consider While Designing CRISPR/Cas Experiments

Designing CRISPR/Cas is easy, as it involves making short DNA target which is unique and is immediately upstream of PAM with less than 4 bp similarity with any other unintended PAM site in the genome. Essentially, while designing SgRNA one has to do a proper genome scan where whole genomic information is available. One important point to remember is that all sgRNAs are not equally effective for gene editing and the reasons behind these remain unknown (please see AddGene e-book, CRISPR 101 as an additional resource available online). Therefore, designing multiple SgRNA depends on the availability of PAM site. In vivo screening of gRNA efficiency is a time-saving strategy, other than that comparing results with other groups/publications/database, which have catalogued sgRNAs (also, refer to Xie et al. 2014). However, GE edited crops, must have a well-established transformation system for the CRISPR/Cas complex to get introduced in the plant cell and editing to take place. Scarcity of PAM may not be an issue where, our aim is to inactivate a gene. However, the scenario changes where precise edits are required which require more efforts (Kleinstiver et al. 2015). The availability of Cas9 orthologs with different PAM sites increases the scope of HDR based precision editing. Moreover, these orthologs do not cross-react and can be used at the same time (Xu et al. 2014). Also, PAM functions as a recognition zone outside 20 bp sequence and does not confer specificity and is not the part of guide RNA. SgRNA must be designed so as to preclude more than 16–17 bp identical to sequence elsewhere in the genome to avoid off-target editing. Another concept related to seed sequence which is defined as the region of 12 bp next to PAM, where there is a maximum probability of editing by NHEJ, that is error-prone repair or random indels. Conserved sequences away from the core or seed sequence also lead to more mismatches and must be taken into consideration while designing experiments. However, with the availability of different PAM sequences recognized by a variant of Cas, the range of sites which can be edited can be increased. Availability of alternate PAM sequences like NGCG, NGAG, and NGAG can overcome the obstacle associated with the limitation imposed by a particular Cas recognition PAM site (Ribeiro et al. 2018). Variants of Cas are obtained from *Staphylococcus aureus* (has NGRRN-PAM site, VRER variant), *S. thermophilus* (NNAGAAW, EQR variant), *Neisseria meningitides* (NNNNGATT, VQR variant) (Kleinstiver et al. 2015). Wang et al. (2016a, b), indicated larger genome size leads to the availability of suitable PAM site in potential target genes. Off-target editing may lead to potential chromosomal rearrangement and unexpected physiological abnormalities in plants. This can be reduced by using sgRNA truncated versions (Fu et al. 2014). The GoldenBraid system is a modular DNA construction tool for synthetic biology work (Vazquez-Vilar et al. 2016) and used in plant genome editing. Here is a list of software and platforms available for CRISPR/Cas work. Tool like CRISPOR (Haeussler et al. 2016), CRISPRscan (Moreno-Mateos et al. 2015), Cas-OFFinder, CasOT, E-CRISPR, CRISPR direct, CCTop, CHOP CHOP, CRISPR-MultiTargeter, CRISPRseek, CasOT, GT Scan can be helpful in designing guide RNA with a minimum off target editing. Baysal et al. (2016) has reported that discrepancy between predicted sgRNA activity by bioinformatic mining and true editing. Wet lab research must be followed to gauze at true editing efficiency. Also, germplasm is prone to natural mutation over time. So, background mutation frequency must be taken into account while taking a call on off-target mutation. Typical CRISPR/Cas workflow in plant GE is shown in Fig. 1.2.

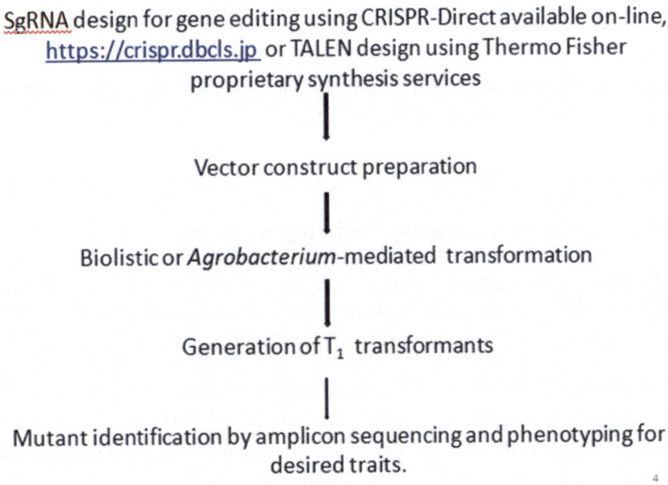

Fig. 1.2 Typical CRISPR/Cas workflow in plant GE

1.6 Practical Application of CRISPR/Cas Technology in Basic and Applied Research

Tang et al. (2017) showed the role of Cpf1 as transcriptional repression in plants. Using the editing system, an introduction of an indel can cause disruption, in the function of the target gene. CRISPR induced sequence replacements help us to modify gene function and modulate expression pattern of gene, which can regulate cell physiology. This can have positive implications in gene regulation. Dead Cas9 (dCas9) with enzyme fusion (is enzymatically active protein domains) will help us to elucidate plant-pathogen dynamics too, helping us design new strategies, for practical application in biotic stress control. Cas9 fusion can be used for epigenetic modification (Hilton et al. 2015). Other application of CRISPR/Cas technology includes gene modulation by base substitution, targeted chromatin modification. Deng et al. (2015) showed the potential application of CRISPR/Cas technology in labelling regions of the chromosome just like DNA-FISH. Gene targeting was also used for targeted recombination between homologous chromosomes. Another technique called enCHIP involves tagging dCas9 with gRNA to specify genes in a given genomic region (Fujita and Fujii 2015). Similarly, TF activation can be performed by targeting histone acetylation or methyl-cytosine demethylases or trimethylases (Benton et al. 2017). Further, m-RNA having alternative sliced sites and polyadenylation of spliceosome surrounding the splice sites at individual introns can aid in controlling gene expression (Jacob and Smith 2017). Practical application of CRISPR/Cas technology will include purification of a gene locus, tag endonuclease proteins and in therapeutic applications (Cox et al. 2015). Also, the GE tool can be used to convert susceptible cultivars to resistance once by rearranging the genetic code as found in the wild type counterpart. Pathogen detection is another area where

CRISPR/Cas can be used, that is detecting a pathogen in crop plants may include Sg-RNA/Cas 13a (C2-C2) based system known as SHERLOCK (Sashital 2018). Cpf1 and Cas 13a can play an important part in translational research particularly pathogen surveillance (Wolter and Puchta 2018).

1.7 Other Types of Genome Editing System

Apart from CRISPR/Cas genome editing, which is the most popular choice, there are also other forms of GE which, will be discussed one by one below (Fig. 1.1b–d).

1.7.1 Meganucleases (MNs) are encoded by mobile introns sometimes known as transposons or selfish genetic elements (see Stoddard 2014 for elaborate details). MNs asymmetric recognition sequence, which is usually 12–40 bp long differentiates itself, from other Type-II restriction enzymes, which have short (4–8 bp) recognition sequence (Guha et al. 2017). MNs occurs infrequently in the plant genome and therefore, are specific to a given sequence, they represent, in most cases. MNs works by using cell's own repair mechanism and therefore, bypassing proofreading mechanism to replicate itself. It is possible to synthesise a range of MNs which are known as hybrid enzymes, with altered protein domain. MNs can recognise these different sequences. One such system is known as Directed Nuclease Editor (DNE) and has been used in agriculturally important cotton crop (website 6). New developments with MNs include the development of MegaTALs (transcription activator-like TALs) effectors, to increase specificity and efficiency. Cellectis holds an exclusive license for the technology (website 7). Modulation of meganuclease target gene activity remains a challenge and therefore, is less popular than other genome editing techniques.

1.7.2 Zinc finger nucleases (ZFNs) are engineered nucleases (bipartite enzymes, 310 amino acid monomer). It has two domains, one is DNA binding and another is cleavage domain, which must dimerise to cleave target DNA sequence (Carrol 2014). This cleavage domain is typically non-specific, type II RE-FokI. The FokI are enzymes found in *Flavobacterium okeanokoites*, which is an endonuclease consisting of DNA binding and cleavage non-specific domain. In short, ZFNs are composed with a series of 3–6 bp finger repeats. The binding domain has cys2-His2 series which can recognize 3 bp. However, designing perfectly binding triplet codon remains a challenge. A minimum of two ZFNs is required to cleave the targeted DNA sequence to initiate HDR (for a detailed review of Zinc nuclease, refer to Durai et al. 2005 and https://www.sangamo.com/technology/genome-editing). Just like Cas9nickase, ZFNickases are available, which has an inactive catalytic domain in one monomer unit, which is required for cleavage (Ramirez et al. 2012). Usages of ZFNs are limited by the time taken to synthesise and their complex assembly process.

1.7.3 TALEN (Transcriptional Activator Like Effector Nucleases) are considered more efficient than ZNFs. TALENs are restriction enzymes engineered to edit

DNA sequence. These are also known by genome editing with engineered nucleases (GEENs). The TALEN's central domain has 13–28 copies of 34 amino acid repeat sequences, with each repeat, targeting individual bases of the DNA sequence (Joung and Sander 2012). Initially, TAL effector molecules were discovered from *Xanthomonas* and are useful for nucleotide recognition. The TAL repeats are identical with two positions, 12–13 AA, which define binding specificity, that is modular DNA binding domain which is fused to the catalytic domain of FokI. One of the proteins bind to the recognition sequence, edit takes place. The endo-nuclease domain of FokI is used to develop ZFN and TALENs. The binding efficiency is high. However, the limitation is imposed by the construction of a new pair of TALs for each target gene. Piatek et al. (2015) reported TALEN can be used to modulate gene expression in *Nicotiana* species with dead Cas9 associated with EDLL and TAL effectors.

The function of ZFNs and TALEN are similar, in sense, that they both create random mutations in absence of a repair template.

1.7.4 Base Editors

Recently, several reports have surfaced related to base editing in plants (Komar et al. Komor et al. 2016; Hua et al. 2018; website 8) which shows the importance of extremely precise editing (C->T and A->G, transition) for certain targets without cut and repair. Thus, base editing has an advantage over traditional CRISPR/Cas based NHEJ and DSB/HR repair, being more precise and predictable. Novel Cas9 fusion which acts as base editors is a fusion combination of Cas with APOBAC1 (rat cytidine deaminase, see Saito et al. 2017). Modified Cas fusion interacts with gRNA to specific base cytosine (C) in the target sequence. Base editing happens in a short window, typically 5 bp from PAM site converting C to T or G to A (Eid et al. 2018). The advantage of using BE is that DNA is not cut and therefore, chances of incorporating non-intended mutation are eliminated. Improvement in BE includes Target-AID base editor (Kuscu and Adli 2016), which has a wider editing window of 15 bp.

1.8 Practical Examples of Genome Editing Application in Crop Plants

As stated earlier in this chapter, crop genome editing involves plant transformation and is an integral part of it. Recalcitrance to transformation is a major problem in crop cultivars. The method of delivery of CRISPR/Cas into cells include *Agrobacterium*-mediated transformation, gene gun, polyethylene glycol (PEG) with protoplast, transfection using cationic polymers, calcium ion-based and lipid-mediated, viral delivery (AAV, Adenovirus, retrovirus, lentivirus) and electro-fusion/ nucleofraction. However, there as several other strategies involved to overcome transformation recalcitrance. For species having low transformation rate, Lowe et al. (2016), showed that Baby boom (*Bbm*) and Wuschel 2 (*Wus* 2) stimulated callus growth in otherwise recalcitrant crop. Taking a cue from such studies, one can edit internal genes (initially having a lower rate of transformation) to establish lines with higher transformation efficiencies. In addition, studies have shown that knock out (KO) of Lig4 and PolQ increases precise repair (Saito et al. 2017) thus, using such lines would invariably increase editing efficiency. The challenge associated with random genome editing (NHEJ) constrains the efficiency of precise homology-directed repair. So, the use of a repair template is advised. Further, an initial round of mitosis may result in mosaic edits. In the G1 phase of cell cycle, HDR is suppressed (Pawluk et al. 2016). We hereby list examples of some of the genome editing crops in agriculture (Table 1.1 and website 9), starting with model species, which were initially used as proof of concept. However, the list of genome-edited plants are increasing day by day and therefore, examples listed here, in this chapter, is by no means exhaustive. There are several Ag companies involved in genome editing with the goal to develop superior

Table 1.1 Typical examples of product developed through genome editing

Sr. No.	Crop	Trait and genes	Regulatory status	Organisation and references
1.	Mushroom	Non-browning white button mushroom developed by inactivating polyphenol oxidase (PPO) gene	No regulation required by USDA	Pennsylvania State University Nature 532, 29 (21 April 2016). https://doi.org/10.1038/nature.2016.19754
2.	Corn	A new and improved waxy corn variety. Deletion in the waxy gene results in essentially eliminating amylose from the kernel	US government would not regulate waxy corn Expected launch, 2020	DuPont-Pioneer https://www.pioneer.com/home/site/us/agronomy/library/crispr-cas/
3.	Canola	Cibus' new SU Canola™ is a non-transgenic (non-GMO) sulfonylurea (SU) herbicide-tolerant canola	Available in USA and Canada	Cibus https://www.cibus.com/products.php
4.	Flax	Cibus' new Flax will be the first non-transgenic (non-GMO) glyphosate-tolerant crop	Expected to launch by 2019	Cibus https://www.cibus.com/products.php
5.	Soybean	High Oleic Soybean Premium Oil	Expected soon	Calyxt http://www.calyxt.com/products/high-oleic-soybean/

products for consumers. Chiefly, Plantedit, Caribou biosciences, Yield10 bioscience, Cibus, Soilcea, Pairwise, Targetgene, 2bladesfoundation, INARI and others (please refer to individual websites for more information).

*1.8.1 Model Plants (*Arabidopsis*, Tobacco, Petunia)*

One of the earliest studies on genome editing in the model plant, *Arabidopsis*, by Feng et al. (2014), found different types of mutation in somatic cells and their inheritance over multiple generations, that is T_1–T_3. Mutation rate ranged from 71 to 79% among T_1–T_3. The commonly occurring mutation found in the study were 1 bp insertions and short deletions. The study showed no evidence of off-target editing. Pyott et al. (2016) showed potyvirus resistance lines in *Arabidopsis* by gene editing. In another study on *Nicotiana tobacum*, another model plant, Gao and colleagues (2015) reported gene editing in *Nt*PDS (phytoene desaturase gene, 81.8% editing), whereas another gene NtPDR6 (pleiotropic drug resistance 6) had 87.5% editing. Here also, authors found no evidence of off-target mutation. Similarly, Nekrasov et al. (2013) earlier had reported targeted editing in *Nicotiana benthamiana* using Cas9-guided endonuclease. Zhang et al. (2016) showed PDS gene editing in *Petunia hybrida* with 55.6–87.5% of T_0 plants showing albino phenotype.

1.8.2 Rice

Yin et al. (2017a, b) showed that knock out of EPFL9 (epidermal patterning factor-like 9) resulted in the reduction of stomata per unit area, thus validating gene function. Wang et al. (2016a, b) have created a KO line of ERF922 (Ethylene responsive factor) for increased blast resistance. Ethylene response regulators are a large family belonging to APETELA 2/ERF super-family in rice which, can be edited to provide enhanced disease resistance without any yield penalty. Shen et al. (2017) reported the role of OsANN3 (rice annexin3 gene) in cold tolerance. Huang et al. 2009 reported the role of a zinc finger protein, DST, which regulates drought and salt tolerance through stomatal aperture regulation. Disruption of OsSWEET14 (sucrose uniporters) promoter by TALEN resulted in disease resistance (Li et al. 2012). Whereas a base pair change in another qSTL3 locus resulted in an increase of stigma length in cv. Kasalth compared to Nipponbare (Li et al. 2015). Ding et al. (2018) showed intron fused to Cpf1, Cas9 sequence in rice showed high efficiency and robustness in targeting multiple sites. This technique can be used to develop enhanced editing tools in plants. Traits of commercial importance like fragrant rice can be obtained by editing OsBADH2 (*betaine aldehyde dehydrogenase)* gene (Shan et al. 2015). Similarly, induced mutations in ALS (Acetolactate synthase) gene resulted in the development of herbicide-resistant non-GM rice (Yu and Powles, 2014). Li et al. (2016) reported four yield-related genes, viz., Gn1a (Grain number), DEP1 (Dense and erect panicle), GS3 (Grain Size), IPA1 (Ideal Plant Architecture 1) as suitable targets for CRISPR/Cas editing. Similarly, Huanga et al. (2018) reported a number of yield genes in rice in several loci. Ma et al. (2018) showed rice blast disease resistance by disrupting OsSEC3A (subunit of the exocyst complex in rice) which also, increased resistance to *Magnaporthe oryzae* with dwarf phenotype. Cheng et al. (2018) showed the function of BIG (involved in auxin transport) gene in plant survival. Xie et al. (2019) reported that editing SPO11 (initiator of meiotic double-stranded breaks), REC8 (meiotic recombination protein), OSD1 (omission of second division), MATL (matrilinear) results in apomictic offspring producer (AOP) line, which is a quadruple mutant. This AOP mutant can be used to propagate hybrids as clonal lines thus, improving production.

1.8.3 Maize

Yield increase was observed in maize over-expressing ARGOS8, an auxin-regulated gene involved in organ size (Shi et al. 2017). Liang et al. (2014) used TALEN and CRISPR/Cas system to edit several genes in maize, that is ZmPDS (Phytoene desaturase), ZmIPK1A, ZmIPK (inositol phosphate kinase), ZmMRP4 (multidrug resistance-associated proteins 4), showing 13–39% editing efficiency. They found both the systems were equally robust in editing maize genes. Char et al. (2017) reported four sgRNA to edit for multiple genes, namely, Argonaute 18 (ZmAgo18a and ZmAgo18b) and dihydroflavonol 4-reductase or anthocyaninless genes (a1 and a4) in maize.

1.8.4 Wheat

There is a global stagnation in wheat yields (see, Zhao et al. 2017 for more information on climate change and yield). Therefore, there is a need to overcome this yield barrier by adopting new plant breeding techniques. Wheat genes have up to three functional versions or even more and therefore, all the active versions when inactivated simultaneously provides an effective phenotype for knock out genes. However, bias towards the gain of function mutation due to easily identifiable phenotype against the recessive ones is masked by gene redundancy and dosage, which delays full utilisation of editing technologies (as reviewed by Borrill et al. 2019). The wheat genome is now sequenced and released by IWGSC. Triticum 3.0 project which assembled 15.3 Gb of data will be used to modify specific gene sequence to produce new mutants. Many more indigenous wheat landraces should also be sequenced and data generated should be used for mining novel source of variation in key genes attributing to yield and disease resistance. Therefore, sources of variation which could be exploited to produce improved varieties can be region-specific natural races, chemically mutagenised population and allelic variation. There are a few reports on GE wheat. Bhowmik et al. (2018) developed a customized WheatCRISPR (https://crispr.bioinfo.nrc.ca/WheatCrispr/) database for designing SgRNAs. They also combined CRISPR/Cas gene-editing system with microspore culture not only to a induce modification in Wheat TaLox2 and TaUbiL1 genes but also got homozygous lines in one generation. Hybrid wheat is a challenge due to inbreeding. Moreover, nuclear MS is not readily available in wheat. Therefore, developing MS sterile lines will help in exploiting hybrid vigour, Whitford et al. (2019). Application of CRISPR/Cas in wheat has remained low due to challenges associated with polyploid genome and transformation system. Howells et al. (2019) targeted PDS (Phytoene desaturase) gene in wheat and obtained edited events with albino phenotypes. The authors also showed TaU6 promoter was more effective than OsU6 promoter, where unfortunately no edits were obtained. They also observed edits which were complex in nature and these edits did not follow Mendelian segregation pattern. The same study also found editing in wheat appears to happen at an early stage. Wang et al. (2018a, b, c) showed multiplexed genome editing involving multiple PAM specific SgRNA which are individually processed. The efficiency of editing ranged from 11 to 17% for single SgRNA guides and 5% for multiple guides. Each of the multiple SgRNA controlled by their own promoters. However, here the size of the CRISPR/Cas9 complex is a limiting factor. Third is utilising Csy4 editor with 28 bp recognition site. Zhang et al. (2018) developed cytidine base-editing in wheat for the target gene ALS (acetolactase deaminase). Base editing offers an advantage as these does not generate double-stranded break. Most of the edits were C to T whereas others had C to G changes. Zhang et al. (2018) targeted common wheat genes (Pinb, waxy, DA1). They used Agrobacterium transformation of protoplast. The mutation frequency in DA1 was 6.8% with editing efficiency of 54.17%. Recently, Humanes et al. (2017) reported high-efficiency gene editing in wheat using a deconstructed version of the wheat dwarf virus (WDV). They obtained 12-fold greater editing at ubiquitin locus than non-linked systems with editing efficiency of nearly 1%. Thus, demonstrating non-integrating complex genome editing in wheat. Plant disease can create havoc, for example, *Puccinia graminis* which caused 90% loss of wheat yield in Uganda. Improvement in GE wheat includes in planta biolistics (iPB) involving the editing of subepidermal cells thus by-passing the regular callus step (Hamada et al. 2017). The same authors edited TaGASR7 gene and obtained 1.4% editing efficiency and edits were inheritable in the next generation. Wang et al. (2014) showed that triple KOs of TaMLO in wheat resulted in inadvertently leaf chlorosis in addition to PM resistance. Editing TaEDR1 in wheat confers PM resistance (Zhang et al. 2017a, b). They also reported that there was no off-target mutation.

1.8.5 Fruit Crops

Recently reported are two studies on genome editing in Banana. Naim et al. (2018) used polycistronic gRNA strategy to edit PDS gene in Cavendish banana cultivar 'William'. The resultant genome-edited plants showed albinism and dwarfism. Since, banana is essentially sterile, GE strategy will help in developing robust disease-resistant lines and novel traits including ones, relevant for industrial applications in near future. Similarly, Kaur et al. (2018) demonstrating CRISPR/Cas editing in PDS gene in banana cv. Rasthali. Peng et al. (2017) reported resistance to citrus canker using CRISPR/Cas9 targeted gene modification of CsLOB1 (*LATERAL ORGAN BOUNDARIES 1*, a family of gene involved in organ development) involved in the promoter region of the gene, in citrus. Other example includes *Phytoplasma* infections in lime which destroyed millions of trees across thousands of orchards could be a potential target for genome editing. Viral infection is another challenging area. The eukaryotic initiation factor is a translation initiation factor (Bastet et al. 2017). EIF4e has an isoform designated iso-EIF4e. The virus infection acts as a lock and key mechanism. EIF4E is essential for translation of RNA to protein for both host plant and pathogen. Therefore, precise editing is needed in the gene of interest to ensure that the physiological state of the crop plant is protected. Mutation in the viral genome linked protein can result in gene redundancy with the recessive EIF4e family (refer, Patrick et al. 2018). Nishitani et al. 2016, reported editing of an endogenous apple phytoene desaturase (*PDS*) gene using the CRISPR/Cas9 system. Malnoy et al. (2016) demonstrated DNA free editing using ribonucleoproteins (RNPs) in apple cv. Golden delicious, by editing *DIPM-1*, *DIPM-2* and *DIPM-4 (alleles of powdery mildew gene)* to increase resistance. Another example is the development of non-browning of artic apple by RNA interference technology (website 11). Similarly, trait can be developed using CRISPR/Cas knock-outs for enhanced disease resistance and also for biotic disease-causing microbes (see, website 12). Wang et al. (2018a, b, c) showed a biallelic mutation in grape with WRKY transcriptional factors which gave resistance to *Botrytis cinerea.*

1.8.6 Oilseeds and Cotton

Jacobs et al. (2015), reported GE in soybean as a proof of concept strategy by knocking out a transgenic line containing GFP and modifying nine other gene loci. Li et al. (2015) reported CRISPR/Cas editing in Soybean at two genomic sites, DD20 (59% editing) and DD43 (76% editing). Further, they were able to mutate a gene, acetolactase synthase 1 (ALS1) at a specific point, P178S, which gave herbicide resistance. Similarly, editing FAD2 (FAD2-1A and FAD2-1B) gene improves oil quality in soybean (Huan et al. 2014). In the homozygous lines, the percentage of oleic acid (component good for health) increased to 80% whereas the linoleic acid component was reduced to 4%. Yang et al. (2017) employed CRISPR/Cas for editing 12 genes in allotetraploid, *Brassica napus* and found editing efficiency for single gene-targeted SgRNA to be around 65.3%. The percentage of mutations which were stably inherited was 48.2 without any reversions. Recently, Yield10 Bioscience reported the development of *Camelina* line which has increased seed oil content and improved quality (website 13). There have been several reports on CRISPR/Cpf1 and Cas9 genome editing in cotton (Li et al. 2019; Zhang et al. 2018). Iqbal et al. (2016) discussed the possibility of using multiplexed CRISPR/Cas for broad-based tolerance against cotton leaf curl virus (CLCuD) and begomovirus complexes. Cotton virus poses a major threat to cotton productivity and thus a GE approach would be well perceived by stakeholders.

1.8.7 Vegetables

Tomato is a model crop to understand gene function due to its ease for genetic transformation (vanEck et al. 2006). Parkhi et al. (2018) reported PDS (Phytoene desaturase) gene editing in tomato parental lines thereby showing a significance of CRISPR/Cas technique in the future commercial development of hybrids. Brooks et al. (2014) edited SlAGO7 (ARGONAUTE) having altered plant phenotype like leaflets without petioles and leaf without lamina. Giannakopoulou et al. (2015) have shown that single nucleotide mutation in a receptor can improve resistance to *Phytoptera* and *Fusarium* in tomato. Cermak et al. (2015) used geminivirus replicons to overcome challenges associated with the efficiency of gene editing for ANT1 involved in anthocyanin in tomato. In another example, tomato plants with KO of MAPK3 (mitogen active protein kinase 3) increased drought tolerance (Ding et al. 2018). Tieman et al. (2017) indicated that flavour component in modern-day tomato cultivar can be improved using biotechnology-related interventions. Similarly, KO of SlAGL6 (slagamous like 6) gene can make tomato fruit parthenocarpic (Klap et al. 2017). Tomelo is a transgene-free tomato resistant to powdery mildew (Nekrasov et al. 2017). Another classical example involve DMR (Downy mildew resistance) 2-oxoglutarate oxygenase mutants, involving disruption of the active site of the enzyme, are partly resistant with an increase in a salicyclic acid (Thomazella et al. 2016). Nekrasov et al. (2017) showed deletion of 48 bp in T_0 plants in tomato, SlMlO1 gene resulting in powdery mildew resistance. Further tested events had no off-target editing. Loss of function MLO allele is associated with yield penalties including spontaneous cell death. Dahan-Meir et al. (2018) targeted carotenoid isomerise (CRISTO), phytoene synthase (PSY1) gene involved in carotenoid biosynthesis pathway using geminiviral replicon and doner template. Studies have also shown that plant viruses can be also used to transmit ZNF nucleases and CRISPR/Cas9 into cells (Zaidi et al. 2016). Nekrasov et al. (2017) targeted PM locus O in tomato for the disease resistance against *Oidium neolycopersici* fungal pathogen. For the vegetative propagated crops, pre-assembled ribo-nucleoproteins (Woo et al. 2015) can be of advantage as such assembled machinery degrades faster leaving virtually no trace of the CRISPR/Cas complex. In another study Lawrenson et al. (2015), used CRISPR/Cas system to edit multi-copy genes in *Brassica oleracea* (BolC.GA4.a) and *Hordeum vulgare* (barley, a field crop) for HvPM19 gene. Potato is often referred as world's most important vegetable crop (Halterman et al. 2015). Since, potato is propagated mostly by vegetative means, and its wild relatives have an abundance of useful genes for insect and pest resistance, which when transferred back to cultivated one can increase yield. Therefore, gene editing can help create a targeted background which will increase the economic value of newly developed varieties. Therefore, potato can be a useful candidate for gene-edited crops. Argentina released PVY-resistant potato, TIC-AR-233-5 (Website 10). The PVY virus can reduce yield by 70% developed by Tecnoplant. Andersson et al. (2016) reported altered starch quality by knocking out the GBSS gene by CRISPR/Cas. The resulting mutants had 1–10 bp indels. The KO-GBSS lines were characterized by phenotypic analysis. Few of the biotech products which are deregulated include INNATE variety of potato. This potato variety is developed by Simplot (website 11). INNATE has reduced black spot, low acrylamide content, low reducing sugar and show resistance to late blight caused by *Phytophthora infestans*. The modified trait was developed by the introduction of the resistance gene from *Solanum venture*. Similarly, CRISPR/Cas strategy can be used to modify endogenous genes and achieve a result on ones obtained by wide-hybridization. Potatoes can be made acrylamide free by knocking out *Ppo5* (Polyphenol oxidase) gene. Viruses pose a serious challenge in crop plants. As reviewed by Roossinck (2015a, b), plant viruses are classified into five major groups including single-stranded DNA viruses belonging to geminivirus (genus begomovirus). These plant viruses are either monopartite or bipartite with a conserved stem-loop sequence having an intergenic region of about 220 bp. CRISPR/Cas approach to virus tolerance involves viral capsid protein (CP), the RCR II motif and the intergenic region (IR). Targeting IR region caused a significant reduction of viral replication and ultimate viral count. The same strategy, however, fails when it comes to edit RNA strands, where Cas13a editing strategy is required. Finally, removing any unwanted mutations or off-target editing, will involve backcrossing which is largely restricted to annual crops with a short life cycle. For detailed deliberation on the application of new plant breeding techniques in Vegetable, please refer to Alamalakala et al. (2016).

1.8.8 Medicinal and Aromatic Plants

In a first CRISPR/Cas based editing in the medicinal plant, opium poppy (*Papaver somniferum* L.), Alagoz et al. (2016) manipulated a gene named OMT2 involved in benzylisoquinoline alkaloids (BIAs) synthesis which is directly related to morphine class of drugs by creating KO lines. This study has relevance in genetically engineering metabolite in non-genome sequenced crops. For deliberation on the application of GE in medicinal plants for secondary metabolite, functional genomics, please see Hu et al. (2016).

1.8.9 Tree Species

Fan et al. (2015) edited *Populus tomentosa* carr. using CRISPR/Cas9. The target gene was phytoene desaturase 8 (PtoPDS) reporting an editing efficiency of 51.7%, showing editing in woody plants. When it comes to perennials or vegetative propagated crops, backcrossing becomes a challenge. Modified Cas system can be used to modify gene expression in perennial plants. An applied application could be increasing yield by manipulating genes involved in photosynthesis efficiency in forest tree species. Perennial plants which have been edited by conventional DNA based GE system, may face challenges in relation to the backcrossing.

1.9 Challenges Associated with CRISPR Edited Crops

The CRISPR/Cas edited crops are bound to face challenges in regards to their commercialization as with any new technology which comes with their own baggage of apprehensions and euphoria. At present GE crops are regulated on the basis of product (Canada) or process (EU) or both (US, on a case-by-case basis, Coordinated Framework for Regulation of Biotechnology). The interpretation of GMO varies between EU member states too, Sweden interprets CRISPR/Cas edited plants as non-GMOs (Wolter and Puchta 2017). Meanwhile, several countries are still contemplating on the issue and yet to come up with guidelines (Ishii and Akari 2017). Experience with the previous generation of biotechnology-derived genetically modified crops, shows a systematic failure to explain science to the consumers and defray ill-informed fear, beforehand (Lucht 2015; Wunderlich and Gatto 2015) especially among European nations. Most consumers in EU have a strong social bias towards conventionally breed crops. This was further aggravated by the fact that existing rigorous regulatory oversight associated with the safety of GM crops led to consumer distrust (Lassoued et al. 2018). Public therefore started perceiving the existing strict regulatory regime in a negative way. Instead of building confidence, scepticism grew. This resulted in negative perception and further fear-mongering by various vested interest groups (especially through mass media) which resulted in the delayed acceptance of biotechnology-derived crops. Therefore, it is imperative that consumers must be appraised and educated in advance for genome-edited products, illustrating, the similarities and contrasting difference between genome-edited crops, from conventionally bred crops, to increase acceptance. Detailed review, by Scheben and Edwards (2018) discuss global patchwork for the future of GE crops. The society at a large must understand these regulations-based technology, process and product-based systems while differentiating biotech-based GM (genetic modified) or GE food products (Marchant and Stevans 2015; Smyth 2017; Tagliabue 2017). This will help us to gather public confidence for crops developed through biotechnology including GE crops. For transgenic crops, Cartagena protocol on biosafety (refer, Hagen and Weiner 1992) was ratified by 170 countries thus arriving at a unified consensus on the regulation of crops produced through the application of biotechnology tools. Regulatory criteria could focus more on food safety, the food itself and not merely on the process employed. The area under biotechnology developed crops, continues to grow steadily in several countries of the world, over the past three decades, ISAAA. Farmers and consumers in some pockets of the world barring EU, have accepted biotech crops regardless of the controversies which spew and are likely to remain so, as the conflict of interest collide among investors in food and Ag sector, in a hypercompetitive world and, food need for the burgeoning population remains a challenge, which needs to be fulfilled. Educating the public regarding different techniques adopted in crop improvement will help them to make informed choices in opting for GE edited food. Political will is a must for GE to succeed. There are various agencies which are involved in regulating various biotechnology-derived products. For example, EU relies on EPSA (European Food Safety Authority) food labelling law. The United States has FDA (Food and Drug Administration Agency), EPA (Environment Protection Agency) and USDA (US Department of Agriculture) and their counterparts in developing and developed world. Stakeholders like farmers, agricultural companies, environmental agencies and activists should be actively engaged in the development of CRISPR/Cas technology and ultimately product dissemination. Improving public perceptions and reducing social objections is must. The first generation of GM crops were produced in soybean, corn, cotton, canola and popularity increased steadily over time after (refer, economic impact of GM crops by Brookes and Barfoot 2014). It is hoped that the genome-edited crops will be seen in a positive light. For developing any biotechnology-derived product there is always high-cost entry barrier for any commercial partner, and therefore delaying approval of genome-edited crops which ultimately lessen its appeal to business and, ultimately the consumer would be left in a lurch. Therefore, it is imperative that timely approval of biotechnology enabled products and a robust regulatory mechanism must exist so as to facilitate the development of a new generation of gene-edited crops. As noted in a recently published WTO communication, thirteen countries including the United States have advocated that undue regulation must not create trade barrier for crops developed through precision breeding agriculture which includes genome-edited crops. The report called for scientifically based harmonisation of regulatory practices (website 14). This is well-timed as EU Court of Justice ruled that GE crops should be regulated in line with GMO, which to many is seen as an overly cautious stand (Callaway 2018). The unwarranted steep regulatory burden puts the additional economic burden to the final product, thus preventing even public players from bringing genome-edited to market. While a considerable amount of taxpayers money is involved in pursuing publically funded research, seldom, agriculture biotech products see the light of the day from funded research. Merely focusing on development process may not help us to realise the full potential of GE. There have been advancements in CRISPR/Cas technology like use of RNP technology involving CRISPR/Cas, which can help us to overcome regulatory over-sight as the CRISPR/Cas RNA protein system remains in the cell for a very short time, thereby ruling out potential off-target editing. Table 1.1 shows some of the DNA-CRISPR/Cas products that have been approved for commercialisation or nearing commercialisation.

1.10 Conclusion

Crop breeding often involves making the best combinations and waiting with the best hope, that the strategy employed, is fulfilled (Bertan et al. 2015). This may require testing thousands of crosses before embarking on the leading path to develop a superior product. Conventional breeding becomes more complicated in polyploidy species, which has multiple copies of a gene. In the era of climate change, there is an urgent need for a second version of the green revolution. Incidence of insect and pest infestations are increasing at an alarming rate and crop lost to biotic factors are causing havoc (Oerke 2018). Thus alternate techniques are required to boost productivity. The first positive step was the gene-edited non-browning mushroom which became the first CRISPR/Cas product to escape regulation in US (Waltz 2016) paving the way for gene-edited products in the market. The mutations induced by genome editing will help to develop new traits in existing or new crop varieties and reduce lengthy conventional breeding timelines. Adoption of transgenic crops faced severe resistance in several pockets of the world, largely, due to consumer distrust, lobbying by vested interest group, bad publicity in the media, a large investment made by companies followed by not so encouraging regulatory regime. Therefore, the public must be educated in advance behind the science used to develop this technology. One way of achieving trust is by undertaking public outreach programs which will ultimately defray environmental and ecological fears surrounding GE crops. To ensure that the genome-edited food is completely safe, looking broadly at the biochemical composition of the resulting food product could be a positive step forward. There needs to have a scientific basis for evaluating safety rather than just basing our verdict on the safety of GE food on a series of presumptions and fear-mongering, that is fear of the unknown and restrictions. As in any sphere of technology, advancement is happening at a rapid pace just like in RNPs sphere. The advantage of using RNPs (pre-assembled CRISPR/Cas9 ribonucleoproteins) is that target action takes place almost immediately and Cas9 protein decomposes by proteases rapidly. Thus, the unintended effect can be reduced by decreasing the activity of Cas9. Further, RNA cannot integrate into the plant DNA. Using RNPs should result in reducing the chances of non-specific mutation and foregone need of the CRISPR assembly to integrate into genome thus no need for further segregation step. Modern genome editing is precise and predictable. The promise RNPs provide in developing consumer-friendly crops could pave way for a more sustainable and environmentally safe agriculture, increasing rapid acceptance of new plant breeding technologies, though the present DNA editing is safe as indicated by-products already launched in the market as presented in Table 1.1. GE edits endogenous genes thus the expectation is that this aspect will help mitigate fear among the general public and increase the acceptability of this technology to sceptics. To conclude, among GE techniques available, CRISPR/Cas is the method of choice, due to its versatility, scalability and cost. Combined with the availability of next-generation sequencing, CRISPR/Cas is the system of choice among GE techniques available with us today. Rephrasing the genetic code will help us to create designer crops and alleviate malnutrition and hunger.

References

Abudayyeh OO, Gootenberg JS, Konermann S et al (2016) C2c2 is a single-component programmable RNA-guided RNA-targeting CRISPR effector. Science 353:557–566
Aglawe SB, Barbadikar KM, Mangrauthia SK, Madhav MS (2008) New breeding technique "genome editing" for crop improvement: applications, potentials and challenges. 3 Biotech 8:336
Alagoz Y, Gurkok T, Zhang B, Unver T (2016) Manipulating the biosynthesis of bioactive compound alkaloids for next-generation metabolic engineering in opium poppy using CRISPR-Cas 9 genome editing technology. Sci Rep 6:309–310
Alamalakala L, Choudhary S, Bhattacharya A, Parkhi V, Mikkileni V, Char BR, Zehr UB (2016) Vegetable breeding through new techniques for higher productivity. In: Chadha KL et al (ed) Doubling farmers income through horticulture. The Horticultural Society of India (Chapter 10), pp 79–84
Anders C, Niewoehner O, Duerst A, Jinek M (2014) Structural basis of PAM-dependent target DNA recognition by the Cas9 endonuclease. Nature 513:569–573
Andersson M, Turesson H, Nicolia A, Fält AS, Samuelsson M, Hof-vander P (2016) Efficient targeted multiallelic mutagenesis in tetraploid potato (*Solanum tuberosum*) by transient CRISPR-Cas9 expression in protoplasts. Plant Cell Rep 36:117–128
Arora L, Narula A (2017) Gene editing and crop improvement using CRISPR-Cas9 system. Front Plant Sci 8:1932
Barrangou R, Marraffini LA (2014) CRISPR-Cas systems: prokaryotes upgrade to adaptive immunity. Mol Cell 54:234–244
Bastet A, Robaglia C, Gallois JC (2017) eIF4E Resistance: natural variation should guide gene editing. Trends Plant Sci 17:S1360–S1385
Baysal C, Bortesi L, Zhu C, Farre G, Schillberg S, Christou P (2016) CRISPR/Cas9 activity in the rice OsBEIIb gene does not induce off-target effects in the closely related paralog OsBE11a. Mol Breed. https://doi.org/10.1007/s11032-016-0533-4
Benton CB, Fiskus W, Bhalla KN (2017) Targeting histone acetylation: readers and writers in Leukemia and Cancer. Cancer 23:286–291
Bertan I, Carvahlo FI, Oliverira AC (2015) Parental selection strategies in plant breeding program. J Crop Biotech 10:211–222
Bhowmik P, Ellison E, Polley B, Bollina V, Kulkarni M, Ghanbarnia K, Song H, Gao C, Voytas DF, Kagale S (2018) Targeted mutagenesis in wheat microspores using CRISPR/Cas9. Sci Rep 8:6502
Boettcher M, McManus MT (2015) Choosing the right tool for the job: RNAi, TALEN or CRISPR. Mol Cell 58:575–585
Bolotin A, Quinquis B, Sorokin A, Ehrlich SD (2005) Clustered regularly interspersed short palindrome repeats (CRISPRs) have spacers of extra-chromosomal origin. Microbiology 151:2551–2561
Borrill P, Harrington SA, Uauy C (2019) Applying the latest advances in genomics and phenomics for trait discovery in polyploid wheat. Plant J 97:56–72
Brookes G, Barfoot P (2014) Economic impact of GM crops. GM Crops Food 5:65–75
Brooks C, Nekrasov V, Lippman ZB, Van Eck J (2014) Efficient gene editing in tomato in the first generation using the clustered regularly interspersed short palindromic repeats/CRISPR-associated 9 system. Plant Physiol 166:1292–1297
Callaway E (2018) CRISPR plants now subject to tough GM laws in European Union. Nature 560:16

Canver MC, Bauer DE, Orkin SH (2017) Functional interrogation of non-coding DNA through CRISPR genome editing. Methods 121:118–129

Carroll D (2014) Genome engineering with zinc-finger nucleases. Genetics 188:773–782

Cermak T, Baltes NJ, Cegan R, Zhang Y, Voytas DF (2015) High frequency, precise modification of the tomato genome. Genome Biol 16:232–246

Char SN, Neelakandan AK, Nahampun H et al (2017) An *Agrobacterium*-delivered CRISPR/Cas9 system for high-frequency targeted mutagenesis in maize. Plant Biotechnol J 15:257–268

Cheng R, Gong L, Zheng S, Liang L (2018) Rice *BIG* gene is required for seedling viability. J Plant Physiol. https://doi.org/10.1016/j.jplph.2018.11.006

Choi PS, Meyerson M (2014) Targeted genomic rearrangements using CRISPR/Cas technology. Nat Commun 5:3728

Cox D, Platt RJ, Zhang F (2015) Therapeutic genome editing: prospects and challenges. Nat Med 21:121–131

Dahan-Meir T, Filler-Hayut S, Melamed-Bessudo Bacobza S, Czosnek H, Aharoni A, Avraham AL (2018) Efficient *in planta* gene targeting in tomato using geminiviral replicons and the CRISPR/Cas9 system. Plant J 95:5–16

Deltcheva E, Chylinski K, Sharma CM, Gonzales K, Chao Y, Pirzada ZA, Eckert MR, Vogel J, Charpentier E (2011) CRISPR RNA maturation by trans-encoded small RNA and host factor RNase III. Nature 471:602–607

Deng W, Shi X, Tjian R, Lionnet T, Singer RH (2015) CASFISH: CRISPR/Cas9-mediated in situ labelling of genomic loci in fixed cells. PNAS 112:11870–11875

Deng D, Chen K, Chen Y, Li H, Xie K (2018) Engineering introns to express RNA guides for Cas9- and Cpf1-mediated multiplex genome editing. Mol Plant 11:542–552

Ding H, He J, Wu Y, Ge C, Wang Y, Zhong S, Peiter E, Liang J, Xu W (2018) The tomato mitogen-activated protein kinase SIMPK1 is a negative regulator of the high temperature stress response. Plant Physiol 177:633–651

Dominguez AA, Lim WA, Qi LS (2016) Beyond editing: repurposing CRISPR–Cas9 for precision genome regulation and interrogation. Nat Rev Mol Cell Biol 17:5–15

Duensing N, Sprink T, Parrott WA, Fedorova M, Lema MA, Wolt JD, Bartsch D (2018) Novel features and considerations for ERA and regulation of crops produced by genome editing. Front Bioeng Biotechnol 6:79

Durai S, Mani M, Kandavelou K, Wu J, Mattew HP, Srinivasan C (2005) Zinc finger nucleases: custom-designed molecular scissors for genome engineering of plant and mammalian cells. Nucleic Acids Res 33:5978–5990

Eid A, Alshareef S, Mahfouz M (2018) CRISPR base editors: genome editing without double-stranded breaks. Biochem J 475:1955–1964

Fan D, Liu TT, Li CF et al (2015) Efficient CRISPR/ Cas9-mediated targeted mutagenesis in *Populus* in the first generation. Sci Rep 5:12217–12223

Feng Z, Mao Y, Xu N et al (2014) Multigeneration analysis reveals the inheritance, specificity, and patterns of CRISPR/Cas-induced gene modifications in *Arabidopsis*. Proc Natl Acad Sci USA 111:4632–4637

Fu Y, Sander JD, Reyon D, Cascio VM, Joung JK (2014) Improving CRISPR-Cas nuclease specificity using truncated guide RNAs. Nat Biotechnol 32:279–284

Fujita T, Fujii H (2015) Isolation of specific genomic regions and identification of associated molecules by engineered DNA-binding molecule-mediated chromatin immunoprecipitation (enChIP) using CRISPR. Methods Mol Biol 1288:43–52

Gao J, Wang G, Ma S et al (2015) CRISPR/Cas9 mediated targeted mutaGenesis in *Nicotiana tabacum*. Plant Mol Biol 87:99–110

Garneau JE, Moineaus (2011) Bacteriophages of lactic acid bacteria and their impact on milk fermentations. Microb Cell Fact 10:S20

Giannakopoulou A, Steele JF, Segretin ME, Bozkurt TO, Zhou J, Robatzek S, Banfield MJ, Pais M, Kamoun S (2015) Tomato I2 immune receptor can be engineered to confer partial resistance

to the Oomycete, *Phytophthora infestans* in addition to the Fungus *Fusarium oxysporum*. Mol Plant Microbe Interact 28:1316–1329
Guha TK, Wai A, Hausner (2017) Programmable genome editing tools and their regulation for efficient genome engineering. Comp Struct Biotechnol J 15:146–160
Haeussler M, Schönig K, Eckert H, Eschstruth A, Mianné J, Renaud J-B, Schneider-Maunoury S, Shkumatava A, Teboul L, Kent J, Joly J-S, Concordet J-P (2016) Evaluation of off-target and on-target scoring algorithms and integration into the guide RNA selection tool CRISPOR. Genome Biol 17:148
Hagen PE, Weiner JB (1992) The Cartagena protocol on biosafety: new rules for international trade in living modified organisms. Convention on biological diversity, U.N. conference on environment and development, June 5, 1992, U.N. Doc. UNEP/Bio.Div/N7-INC.S/4
Halterman D, Guenthner J, Collinge S, Butler N, Douches D (2015) Biotech potatoes in the 21st century: 20 years since the first biotech potato. Am J Potato 93:1–20
Hamada H, Linghu Q, Nagira Y, Miki R, Taoka N, Imai R (2017) An in planta biolistic method for stable wheat transformation. Sci Reports 7:11443
Hille F, Charpentier E (2016) CRISPR-Cas: biology, mechanisms and relevance. Philos Trans R Soc Lond B Biol Sci 5:371
Hilton IB, D'Ippolito AM, Vockley CM, Thakore PI, Crawford GE, Reddy TE (2015) Epigenome editing by a CRISPR-Cas9-based acetyltransferase activates genes from promoters and enhancers. Nat Biotechnol 33:510–517
Howells RM, Craze M, Bowden S, Wallington EJ (2019) Efficient generation of stable, heritable gene edits in wheat using CRISPR/Cas9. BMC Plant Biol 18:215
Hsu PD, Lander ES, Zhang F (2015) Development and applications of CRISPR-Cas9 for genome engineering. Cell 157:1262–1278
Hu TY, Gao W, Huang LQ (2016) Prospecting application of CRISPR/Cas9 genome editing technology in research of medicinal plants. China J Chin Materia Med 41(16):2953–2957
Hua K, Tao X, Yuan F, Wang D, Zhu JK (2018) Precise A•T to G•C base editing in the rice genome. Mol Plant 11:627–630
Huai C, Li G, Yao R, Zhang Y, Cao M, Kong L, Jia C, Yuan C, Chen H, Lu D, Huang Q (2017) Structural insights into DNA cleavage activation of CRISPR-Cas9 system. Nat Commun 8:1375
Huan W, Coffman A, Clasen BM, Demorest ZL, Lowy A, Ray E et al (2014) Improved soybean oil quality by targeted mutagenesis of the fatty acid desaturase 2 gene family. Plant Biotech J 12:934–940
Huang XY, Chao DY, Gao JP, Zhu MZ, Shi M, Lin HX (2009) A previously unknown zinc finger protein, DST, regulates drought and salt tolerance in rice via stomatal aperture control. Genes Dev 23:1805–1817
Huanga J, Lia J, Zhoub J, Wanga L, Yanga S, Hurstc LD, Li WH, Tiana H (2018) Identifying a large number of high-yield genes in rice by pedigree analysis, whole-genome sequencing, and CRISPR-Cas9 gene knockout. PNAS 115:E7559–E7567
Humanes JG, Wang Y, Liang Z, Shan Q, Ozuna CV, Saìnchez-Leoìn S (2017) High-efficiency gene targeting in hexaploid wheat using DNA replicons and CRISPR/Cas9. Plant J 89:1251–1262
Iqbal Z, Sattar MN, Shafiq M (2016) CRISPR/Cas9: a tool to circumscribe cotton leaf curl disease. Front Plant Sci 7:475
Ishii T, Araki M (2017) A future scenario of the global regulatory landscape regarding genome-edited crops. GM Crop Food 8:44–56
Ishino Y, Krupovic M, Forterre P (2018) History of CRISPR-Cas from encounter with a mysterious repeated sequence to genome editing technology. J Bacteriology. https://doi.org/10.1128/JB.00580-17
Jacob AG, Smith CWJ (2017) Intron retention as a component of regulated gene expression programs. Hum Genet 136:1043–1057
Jacobs TB, LaFayette PR, Schmitz RJ, Parrott WA (2015) Targeted genome modifications in soybean with CRISPR/Cas9. BMC Biotechnol 15:16. https://doi.org/10.1186/s12896-015-0131-2

Jinek M, Chylinski K, Fonfara I, Hauer M, Doudna JA, Charpentier E (2012) A programmable dual-RNA guided DNA endonuclease in adaptive bacterial immunity. Science 337:816–821

Joung JK, Sander JD (2012) TALENs: a widely applicable technology for targeted genome editing. Nat Rev Mol Cell Biol 14:49–55

Karginov FV, Hannon GJ (2010) The CRISPR system: small RNA-guided defense in bacteria and archaea. Mol Cell 37:7

Kaur N, Alok A, Shivani Kaur N, Pandey P, Awasthi P, Tiwari S (2018) CRISPR/Cas9-mediated efficient editing in phytoene desaturase (PDS) demonstrates precise manipulation in banana cv. Rasthali genome. Funct Integr Genomics 18:89–99

Kimberland ML, Hou W, Alfonso-Pecchio A, Wilsons S, Rao Y, Zhang S, Lu Q (2018) Strategies for controlling CRISPR/Cas9 off-target effects and biological variations in mammalian genome editing experiments. J Biotechnol 284:91–101

Klap C, Yeshayahou E, Bolger AM, et al (2017) Tomato facultative parthenocarpy results from SlAGAMOUS-LIKE 6 loss of function. Plant Biotechnol J 15:634–647

Kleinstiver BP, Prew MS, Tsai SQ et al (2015) Engineered CRISPR-Cas9 nucleases with altered PAM specificities. Nature 523:481–485

Komor AC, Kim YB, Packer MS, Zuris JA, Liu DR (2016) Programmable editing of a target base in genomic DNA without double-stranded DNA cleavage. Nature 533:420–424

Kuscu C, Adli M (2016) CRISPR-Cas9-AID base editor is a powerful gain-of-function screening tool. Nat Methods 13:983–984

Larson MH, Gilbert LA, Wang X, Lim WA, Weissman JS, Qi LS (2013) CRISPR interference (CRISPRi) for sequence-specific control of gene expression. Nat Protoc 8:2180–2196

Lassoued R, Smyth SJ, Phillips PWB, Hesseln H (2018) Regulatory uncertainty around new breeding techniques. Front Plant Sci 9:1291

Lawrenson T, Shorinola O, Stacey N et al (2015) Induction of targeted, heritable mutations in barley and *Brassica oleracea* using RNA-guided Cas9 nuclease. Genome Biol 16:258–271

Li T, Liu B, Spalding MH, Weeks DP, Yang B (2012) High-efficiency TALEN-based gene editing produces disease-resistant rice. Nat Biotechnol 30:390–392

Li Z, Liu Z, Xing A, Moon BP, Koellhoffer JP, Huang L (2015) Cas9-guide RNA directed genome editing in soybean. Plant Physiol 169:960–970

Li M, Li X, Zhou Z, Wu P, Fang M, Pan X, Lin Q, Luo W, Wu G, Li H (2016) Reassessment of the four yield-related genes *Gn1a*, *DEP1*, *GS3*, and *IPA1* in rice using a CRISPR/Cas9 system. Front Plant Sci 7:377

Li L, Wei K, Zheng G, Liu X, Chen S, Jiang W, Lu Y (2018a) CRISPR-Cpf1-assisted multiplex genome editing and transcriptional repression in streptomyces. Appl Environ Microbiol 31:84

Li X, Wang Y, Chen S, Tian H, Fu D, Zhu B, Luo Y, Zhu H (2018b) Lycopene is enriched in tomato fruit by CRISPR/Cas9-mediated multiplex genome editing. Front Plant Sci 9:559

Li X, Wang Y, Liu Y, Yang B et al (2018c) Base editing with a Cpf1-cytidine deaminase fusion. Nat Biotechnol 36:324–327

Li B, Rui H, Li Y, Wang Q, Alariqi M, Qin L, Sun L, Ding X, Wang F, Zou J, Wang Y, Yuan D, Zhang X, Jin S (2019) Robust CRISPR/Cpf1(Cas12a) mediated genome editing in allotetraploid cotton (G. hirsutum). Plant Biotech J. https://doi.org/10.1111/pbi.13147

Liang Z, Zhang K, Chen K, Gao C (2014) Targeted mutagenesis in *Zea mays* using TALENs and the CRISPR/Cas system. J Genet Genomics 41:63–68

Lowe K, Wu E, Wang N, Hoerster G, Hastings C (2016) Morphogenic regulators baby boom and wuschel improve monocot transformation. Plant Cell 28:1998–2015

Lucht JM (2015) Public acceptance of plant biotechnology and GM Crops. Viruses 7:4254–4281

Ma X, Zhang Q, Zhu Q et al (2015) A robust CRISPR/Cas9 system for convenient, high-efficiency multiplex genome editing in monocot and dicot plants. Mol Plant 8:1274–1284

Ma J, Chen J, Wang M, Ren Y, Wang S, Lei C, Chen Z, Sodmegen A (2018) Disruption of OsSEC3A increases the content of salicylic acid and induces plant defense responses in rice. J Exp Bot 69:1051–1064

Makarova KS, Koonin EV (2015) Annotation and classification of CRISPR-Cas systems methods. Mol Biol 1311:47–75
Makarova KS, Wolf YI, Koonin EV (2018) Classification and nomenclature of CRISPR-Cas systems: where from here? CRISPR J 1:5
Malnoy M, Viola R, Jung MH et al (2016) DNA-free genetically edited grapevine and apple protoplast using CRISPR/Cas9 ribonucleoproteins. Front Plant Sci 7:1904
Marchant GE, Stevans YA (2015) A new window of opportunity to reject process-based biotechnology regulation. GM Crops & Food 6(4):233–242
Marco MB, Moineau S, Quiberoni A (2012) Bacteriophages and dairy fermentations. Bacteriophage 2:149–158
Mohanta TP, Bashir T, Harshem A, Fathi AE, Bae H (2017) Genome editing tools in plants. Genes (Basel) 8:399
Mojica FJM, Rodriguez-Valera (2016) The discovery of CRISPR in archaea and bacteria. The FEBS J 283:3162–3169
Mojica FJ, Ferrer C, Juez G, Rodriguez-Valera F (1995) Long stretches of short tandem repeats are present in the largest replicons of the *Archaea*, *Haloferax mediterranei* and *Haloferax volcanii* and could be involved in replicon partitioning. Mol Microbiol 17:85–93
Moreno-Mateos MA, Vejnar CE, Beaudoin J, Fernandez JP, Mis EK, Khokha MK, Giraldez AJ (2015) CRISPRscan: designing highly efficient sgRNAs for CRISPR-**Cas9 targeting** *in vivo*. Nat Methods 12:982–988
Murovec J, Pirc Z, Yang B (2017) New variants of CRISPR RNA-guided genome editing enzymes. Plant Biotechnol J 15:917–926
Naim F, Dugdale B, Kleidon J, Brinin A, Shand K, Waterhouse P, Dale J (2018) Gene editing the phytoene desaturase alleles of Cavendish banana using CRISPR/Cas9. Transgenic Res. https://doi.org/10.1007/s11248-018-0083-0
Nekrasov V, Staskawicz B, Weigel D, Jones JD, Kamoun S (2013) Targeted mutagenesis in the model plant *Nicotiana benthamiana* using Cas9-guided endonuclease. Nat Biotechnol 31:691–693
Nekrasov V, Wang C, Win J, Lanz C, Weigel D, Kamoun S (2017) Rapid generation of a transgene-free powdery mildew resistant tomato by genome deletion. Sci Reports 7:482
Nishitani C, Hirai N, Komori S et al (2016) Efficient genome editing in apple using a CRISPR/Cas9 system. Sci Rep 6:314–381
Oerke EC (2006) Crop losses to pests. J Agric Sci 144:31–34
Parkhi V, Bhattacharya A, Choudhary S, Pathak R, Gawade V, Palan B, Alakalamala L, Mikkilineni V, Char BR (2018) Demonstration of CRISPR- cas 9-mediated pds gene editing in a tomato hybrid parental line. Ind J Gen Plant Breed 78:132
Patrick RM, Lee JCH, Teetsel JRJ, Yang SH, Choy GM, Browning KS (2018) Discovery and characterization of conserved binding of eIF4E 1 (CBE1), a eukaryotic translation initiation factor 4E–binding plant protein. https://doi.org/10.1074/jbc.ra118.003945
Pawluk A, Amrani N, Zhang Y, Garcia B, Hidalgo-Reyes Y, Lee J, Edraki A, Shah M, Sonthemier EJ, Maxwell KL, Davidson AR (2016) Naturally occurring off-switches for CRISPR-Cas9. Cell 167:1–10
Pawluk A, Davidson AR, Maxwell KL (2018) Anti-CRISPR: discovery, mechanism and function. Nat Rev Microbiol 16:12–17
Peng A, Chen S, Lei T et al (2017). Engineering canker-resistant plants through CRISPR/Cas9-targeted editing of the susceptibility gene CsLOB1 promoter in citrus. Plant Biotechnol J. https://doi.org/10.1111/pbi.12733
Piatek A, Ali Z, Baazim H, Li L, Abulfaraj A, Al-Shareef S, Aouida M et al (2015) RNA-guided transcriptional regulation in planta via synthetic dCas9-based transcription factors. Plant Biotechnol J 13:578–589
Pyott DE, Sheehan E, Molnar A (2016) Engineering of CRISPR/ Cas9-mediated potyvirus resistance in transgene-free *Arabidopsis* plants. Mol Plant Pathol 17:1276–1288

Qi LS, Larson MH, Gilbert LA, Doudna JA, Weissman JS, Arkin AP, Lim WA (2013) Repurposing CRISPR as an RNA-guided platform for sequence-specific control of gene expression. Cell 152:1173–1183

Ramirez CL, Certo MT, Mussolino C, Goodwin MJ, Cradick TJ, McCaffrey AP, Cathomen T, Scharenberg AM, Joung JK (2012) Engineered zinc finger nickases induce homology-directed repair with reduced mutagenic effects. Nucleic Acid Res 40:5560–5568

Ribeiro LF, Ribeiro LFC, Barreto MQ, Ward RJ (2018) Protein engineering strategies to expand CRISPR-Cas9 applications. Int J Genom. Article ID-1652567

Roossinck M (2015a) Metagenomics of plant and fungal viruses reveals an abundance of persistent lifestyles. Front Microbiol 5:767

Roossinck MJ (2015b) Plants, viruses and the environment: ecology and mutualism. Virology 480:271–277

Ross-Ibarra J, Morrell P, Gaut BS (2007) Plant domestication, a unique opportunity to identify the genetic basis of adaptation. PNAS 104:8641–8648

Saito S, Maeda R, Adachi N (2017) Dual loss of human *POLQ* and *LIG4* abolishes random integration. Nat Commun 8:16112

Sander JD, Joung (2014) CRISPR-Cas systems for genome editing, regulation and targeting. Nat Biotechnol 32:347–355

Sashital DG (2018) Pathogen detection in the CRISPR-Cas era. Genome Med 10:32

Schaart JG, Weil CC, Lotz LAP, Smulders MJM (2016) Opportunities for products of new plant breeding techniques. Trends Pl Sci 21:438–449

Scheben A, Edwards D (2018) Bottlenecks for genome-edited crops on the road from lab to farm. Genome Biol 19:178

Schiml S, Fauser F, Puchta H (2014) The CRISPR/Cas9 system can be used as nuclease for in planta gene targeting and as paired nickases for directed mutagenesis in Arabidopsis resulting in heritable progeny. Plant J 88:1139–1150

Sekine R, Kawata T, Muramoto T (2018) CRISPR/Cas9 can be used to simultaneously edit several genes. Sci Rep 8:8471

Seol JH, Shim EY, Lee SE (2018) Microhomology-mediated end joining: good, bad and ugly. Mutat Res 809:81–87

Shabbir MAB, Hao H, Shabbir MZ, Hussain HI, Iqbal Z, Ahmed S, Sattar A, Iqbal M, Li J, Yuan Z (2016) Survival and evolution of CRISPR–Cas system in prokaryotes and its applications. Front Immunol 7:375

Shan Q, Zhang Y, Chen K, Zhang K, Gao C (2015) Creation of fragrant rice by targeted knockout of the OsBADH2 gene using TALEN technology. Plant Biotechnol J 13:791–800

Shen C, Que Z, Xia Y, Tang N, Li D, He R, Cao M (2017) Knock out of annexin gene OsAnn3 via CRISPR/Cas9-mediated genome editing decreased cold tolerance in rice. J Plant Biol 60:539–547

Shi J, Gao H, Wang H et al (2017) ARGOS8 variants generated by CRISPR-Cas9 improve maize grain yield under field drought stress conditions. Plant Biotechnol J 16:299–311

Slaymaker IM, Gao L, Zetsche B, Scott DA, Yan WX, Zhang F (2016) Rationally engineered Cas9 nucleases with improved specificity. Science 351:84–88

Smykal P, Nelson MN, Berger JD, Wettberg EJB (2018) The impact of genetic changes during crop domestication. Agronomy 8:19

Smyth SJ (2017) Canadian regulatory perspectives on genome engineered crops. GM Crops Food 8:35–43

Songstad DD, Petolino JF, Voytas DF, Reichert NA (2017) Genome editing of plants. Critical Rev Plant Sci 36:1–23

Stoddard BL (2014) Homing endonucleases from mobile group I introns: discovery to genome engineering. Mob DNA 5:7

Sundaresan R, Parameshwaran HP, Yogesha SD, Keilbarth MW, Ranjan R (2017) RNA-independent DNA cleavage activities of Cas9 and Cas12a. Cell Rep 21:3728–3739

Tagliabue G (2017) Product, not process! Explaining a basic concept in agricultural biotechnologies and food safety. Life Sci Soc Policy 13:3

Tang X, Lowder LG, Zhang T, Malzahn AA, Zheng X, Voytas DF, Zhong Z, Chen Y, Ren Q, Li Q, Kirkland ER, Zhang Y, Qi Y (2017) A CRISPR-Cpf1 system for efficient genome editing and transcriptional repression in plants. Nat Plants 3:1701

Thomazella D, Brail Q, Dahlbeck D, Staskawic Z (2016) CRISPR-Cas9 mediated mutagenesis of a DMR6 ortholog in tomato confers broad-spectrum disease resistance. bioRxiv preprint. https://doi.org/10.1101/064824

Tieman D, Zhu G, Resende MF Jr, Lin T, Nguyen C, Bies D (2017) A chemical genetic roadmap to improved tomato flavor. Science 355:391–394

Tycko J, Myer VE, Hsu PD (2017) Methods for optimizing CRISPR-Cas9 genome editing specificity. Mol Cell 63:355–370

Unniyampurath U, Pilankatta R, Krishnan MN (2016) RNA interference in the age of CRISPR: will CRISPR interfere with RNAi? Int J Mol Sci 17:291

Van Eck J, Kirk DD, Walmsley AM (2006) Tomato (*Lycopersicum esculentum*). Methods Mol Biol 343:459–473

Vaughan DA, Balazs E, Heslop-Harrison JS (2017) From crop domestication to super-domestication. Ann Bot 100:893–901

Vazquez-Vilar M, Bernabe-Orts JM, Fernandez-Del-Carmen A, Ziarsolo P, Blanca J, Granell A (2016) A modular toolbox for gRNA-Cas9 genome engineering in plants based on the golden braid standard. Plant Methods 12:10–21

Verwaal R, Buiting-Wiessenhann N, Dalhuijsen S, Roubos JA (2017) CRISPR/Cpf1 enables fast and simple genome editing of *Saccharomyces cerevisiae*. Yeast 35:2

Waltz E (2016) Gene-edited CRISPR mushroom escapes US regulation. Nature 532:293

Wang Y, Cheng X, Shan Q, Zhang Y, Liu J, Gao C (2014) Simultaneous editing of three homoeoalleles in hexaploid bread wheat confers heritable resistance to powdery mildew. Nat Biotechnol 32:947–951

Wang F, Wang C, Liu P et al (2016a) Enhanced rice blast resistance by CRISPR/Cas9-targeted mutagenesis of the ERF transcription factor gene OsERF922. PLoS ONE 11:e0154027

Wang Y, Liu X, Ren C, Zhong GY, Yang L, Li S, Liang Z (2016b) Identification of genomic sites for CRISPR/Cas9-based genome editing in the *Vitis vinifera* genome. BMC Plant Biol 16:96

Wang X, Tu M, Wang D, Liu J, Li Y, Li Z, Wang Y, Wang X (2018a) CRISPR/Cas9-mediated efficient targeted mutagenesis in grape in the first generation. Plant Biotech 16:844–855

Wang W, Simmonds J, Pan Q, Davidson D, He F, Battal A, Akhunova A, Trick HN, Uauy C, Akhunov E (2018b) Gene editing and mutagenesis reveal inter-cultivar differences and additivity in the contribution of TaGW2 homoeologues to grain size and weight in wheat. Theor Appl Genet 131:2463–2475

Wang W, Pan Q, Akhunova A, Chao S, Trick H, Akhunov E (2018c) Transgenerational CRISPR-Cas9 activity facilitates multiplex gene editing in allopolyploid wheat. CRISPR J 1:65–74

Wolter F, Puchta H (2017) Genome engineering mit CRISPR/Cas — Revolution in der Pflanzenzüchtung. Biospektrum 23:159–161

Wolter F, Puchta H (2018) The CRISPR/Cas revolution reaches the RNA world: Cas13, a new Swiss Army knife for plant biologists. Plant J 94:767–775

Woo JW, Kim J, Kwon SI et al (2015) DNA-free genome editing in plants with pre-assembled CRISPR-Cas9 ribonucleoproteins. Nat Biotechnol 33:1162–1164

Wunderlich S, Gatto KA (2015) Consumer perception of genetically modified organisms and sources of information. Adv Nutr 6:842–851

Xie K, Yang Y (2013) RNA-guided genome editing in plants using a CRISPR/Cas system. Mol Plant 6:1975–1983

Xie K, Zhang J, Yang Y (2014) Genome-wide prediction of highly specific guide RNA spacers for the CRISPR/Cas9 mediated genome editing in model plants and major crops. Mol Plant 7:923–926

Xie E, Li Y, Tang D, Lu Y, Shen Y, Cheng Z (2019) A strategy for generating rice apomixes by gene editing. J Plant Integr Biol. https://doi.org/10.1111/jipb.12785

Xu T, Li Y, Van NJD, He Z, Zhou J (2014) Cas9-based tools for targeted genome editing and transcriptional control. App Environ Microbiol 80:1544–1552

Yang H, Wu JJ, Tang T, Liu KD, Dai C (2017) CRISPR/Cas9-mediated genome editing efficiently creates specific mutations at multiple loci using one sgRNA in *Brassica napus*. Sci Rep 8:7489

Yin K, Gao C, Qiu JL (2017a) Progress and prospects in plant genome editing. Nat Plants 3:17107

Yin X, Biswal AK, Dionora JH, Perdigon KM, Balahadia CP, Mazumdar S, Chater C, Lin HC, Coe RA, Kretzschmar T, Gray JE, Quick PW, Bandyopadhyay A (2017b) CRISPR-Cas9 and CRISPR-Cpf1 mediated targeting of a stomatal developmental gene EPFL9 in rice. Plant Cell Rep 36:745–757

Yoshimi K, Kunihiro Y, Kaneko T, Nagahora H, Voigt B, Mashimo T (2016) ssODN-mediated knock-in with CRISPR-Cas for large genomic regions in zygotes. Nat Commun 7:1043

Yu Q, Powles SB (2014) Resistance to AHAS inhibitor herbicides: current understanding. Pest Manag Sci 70:1340–1350

Zaidi SS, Tashkandi M, Mansoor S, Mahfouz MM (2016) engineering plant immunity: using CRISPR/Cas9 to generate virus resistance. Front Plant Sci 7:1673

Zetsche B, Gootenberg JS, Abudayyeh OO, Slaymaker IM, Makarova KS, Essletzbichler P, Volz SE, Joung J, van der Oost J, Regev A, Koonin EV, Zhang F (2015) Cpf1 is a single RNA-guided endonuclease of a class 2 CRISPR-Cas system. Cell 163:759–771

Zhang B, Yang X, Yang C, Li M, Guo Y (2016) Exploiting the CRISPR/Cas9 system for targeted genome mutagenesis in petunia. Sci Rep 6:20315

Zhang J, Wang X, Yao J, Li Q, Lui F, Yotsukara N, Krupnova TN, Duan D (2017a) Effect of domestication on the genetic diversity and structure of *Saccharina japonica* populations in China. Sci Rep 7:42158

Zhang Y, Bai Y, Wu G, Zou S, Chen Y, Gao C, Tang D (2017b) Simultaneous modification of three homoeologs of *TaEDR1* by genome editing enhances powdery mildew resistance in wheat. https://doi.org/10.1111/tpj.13599

Zhang Z, Ge X, Luo X, Wang P, Qiang F, Hu G, Xiao J, Li Fuguang WuJ (2018) Simultaneous editing of two copies of *Gh14-3-3d* confers enhanced transgene-clean plant defense against *Verticillium dahliae* in allotetraploid upland cotton. Front Plant Sci 9:842

Zhao C, Liu B, Piao S, Wang X, Lobell DB, Huang Y, Huang M, Yao Y, Bassu S, Ciais P, Durand JL, Elliott J, Ewert F, Janssens IA, Li T, Lin E, Liu Q, Martre P, Müller C, Peng S, Peñuelas J, Ruane AC, Wallach D, Wang T, Wu D, Liu Z, Zhu Y, Zhu Z, Asseng S (2017) Temperature increase reduces global yields of major crops in four independent estimates. Proc Natl Acad Sci USA 114:9326–9331

Websites

Website 1. fao.org/save-food/resources/keyfindings/en/. Accessed on 15 Nov 2018

Website 2. http://www.un.org/en/development/desa/news/population/2015-report.html

Website 3. FAO/IAEA (2014) Plant breeding and genetics. Food and Agricultural Organization, International Atomic Energy Agency Division of Nuclear Techniques in Agriculture (http://www-naweb.iaea.org/nafa/pbg/). Accessed on 15 Nov 2018

Website 4. There's CRISPR in Your Yogurt. https://www.the-scientist.com/notebook/theres-crispr-in-your-yogurt-36142. Accessed 15 Nov 2018

Website 5. https://www.sciencemag.org/news/2015/12/and-science-s-2015-breakthrough-year. Accessed on 15 Nov 2018

Website 6. https://www.businesswire.com/news/home/20121126005300/en/Precision-BioSciences-Announces-Publication-Cotton-Trait-Stacking. Accessed 15 Nov 2018

Website 7. https://news.cision.com/bmc-press-room/r/cellectis-acquires-exclusive-license-to-tal-effector-patents-from-university-of-minnesota,g537967. Accessed 15 Nov 2018

Website 8. https://www.technologyreview.com/s/609203/crispr-20-is-here-and-its-way-more-pre cise/ (base editing). Accessed 15 Nov 2018
Website 9. https://www.nationalgeographic.com/environment/future-of-food/food-technology-gene-editing/. Accessed 15 Nov 2018
Website 10. https://www.freshfruitportal.com/news/2018/08/21/argentina-approves-disease-resist ant-gm-potato/. Accessed on 15 Nov 2018
Website 11. https://www.livescience.com/48870-genetically-engineered-arctic-apple.html. Accessed on 15 Nov 2018
Website 12. https://www.technologyreview.com/s/600765/10-breakthrough-technologies-2016-pre cise-gene-editing-in-plants/ (plant pathogens) Accessed on 15 Nov 2018
Website 13. https://globenewswire.com/news-release/2018/10/02/1588493/0/en/Yield10-Biosci ence-Obtains-Nonregulated-Status-for-its-Novel-CRISPR-Cas9-Triple-Gene-Edited-Camelina-Plant-Lines-to-Boost-Seed-Oil-Content.html
Website 14. https://www.state.gov/r/pa/prs/ps/2018/11/287097.htm. International Statement on Agricultural Applications of Precision Biotechnology. Accessed 15 Nov 2018

Chapter 2
Application of Bioinformatics Tools in CRISPR/Cas

Shalu Choudhary, Abhijit Ubale, Jayendra Padiya, and Venugopal Mikkilineni

Abstract Clustered regularly interspaced short palindromic repeats (CRISPR) target sequence selection is a key step for achieving successful targeted gene editing. Possible off-target editing and variance in editing efficacy imposes restrictions on target site selection. In recent years, for conducting CRISPR knock-in/out experiments, a plethora of computational tools have been developed to aid researchers in target site selection. These tools are likely to help in selecting the target site and in designing highly specific single-guide RNA (sgRNA) for CRISPR applications. The broadening of CRISPR applications has also seen the development of sgRNA design tools specifically for CRISPR-mediated gene regulation and genetic screening studies. Besides assisting in sgRNA design, computational tools have been developed to evaluate CRISPR genome-edited data generated through Next Generation Sequencing (NGS) platforms. The computational tools developed have their own salient features, differ in design specifications, input parameters, genomes, output data and visualisation. Therefore, it is important for researchers to choose the appropriate tools suitable for their experimental design. In this chapter, we have comprehensively outlined the various sgRNA design tools available in plants for use in different CRISPR applications and their on-target efficiency prediction models and off-target detection. These models provide us an insight into the underlying criteria used while developing CRISPR tools. We also discuss computational tools developed for genome editing assessment using NGS data. This chapter walks the reader through a collection of computational tools along with their features available for CRISPR applications, especially in the context of plants. We hope that this compiled information on CRISPR bioinformatics tools would benefit plant scientists while performing their CRISPR/Cas experiment.

Keywords CRISPR/Cas9 · Gene editing · sgRNA design · On-target activity · Off-target sites · PAM · Efficacy and specificity

S. Choudhary · A. Ubale · J. Padiya · V. Mikkilineni (✉)
Mahyco Private Limited, Aurangabad-Jalna Road, Dawalwadi, Jalna, Maharashtra, India
e-mail: venugopal.mikkilineni@mahyco.com

A. Bhattacharya et al. (eds.), *CRISPR/Cas Genome Editing*, Concepts and Strategies in Plant Sciences, https://doi.org/10.1007/978-3-030-42022-2_2

2.1 Introduction

The three foremost genome editing technologies, Transcription activator-like effector nucleases (TALENs), Zinc finger nucleases (ZFNs) and CRISPR/Cas9, use programmable nucleases to modify a specific region of the genome in a precise and predictable manner. While TALENs and ZFNs, depend on protein-based system, CRISPR/Cas9 technology uses RNA guided nucleases. As a result, in a short span, CRISPR/Cas9 technology has been established as a powerful approach for the precise introduction of DNA double-strand break (DSB) in plant genome across diverse plant species (Belhaj et al. 2013; Bortesi and Fischer 2015; Jung et al. 2017). CRISPR application to induce targeted knock-out of the gene of interest has been deployed for the development of new cultivars with improved novel traits (Jung et al. 2017; Hussain et al. 2018; Yin and Qiu 2019). Recently, researchers have been adopting CRISPR/Cas9 system as a tool for genetic screening discovering gene function. Studies in rice (Meng et al. 2017; Lu et al. 2017) and tomato (Jacobs et al. 2017) have demonstrated the feasibility of generating CRISPR/Cas9 mutant libraries. Further, applications of CRISPR technology is constantly advancing to engineer other kinds of genetic manipulations in plant systems such as for gene knock-in (Li et al. 2013; Collonnier et al. 2017), multiplexed gene editing (Wang et al. 2018; Hashimoto et al. 2018) and targeted gene inhibition or activation (CRISPRi/a) (Piatek et al. 2015; Park et al. 2017b). In all, the simplicity, robustness and versatility of CRISPR/Cas system offer great potential for enhancement of targeted traits in plants that could boost crop breeding.

From the genetic engineering perspective, Type II CRISPR/Cas9 system adapted from the bacterium, *Streptococcus pyogenes*, consists of two parts: chimeric sgRNA and Cas9 endonuclease. The sgRNA forms a functional complex with Cas9 and guides the nuclease to cleave target DNA sequence by locating the protospacer adjacent motif (PAM) sequence that varies with the type of Cas9 protein used. The most commonly used Cas9 to date is SpCas9 with PAM 5'-NGG-3'. The Cas9 protein will cleave the sequence only in the presence of PAM. sgRNA made in vitro or in vivo consists of custom-designed crRNA (CRISPR RNA) sequence fused to the tracrRNA (trans-activating crRNA) sequence. The crRNA sequence is a 5′-end 20-nt variable part known as gRNA spacer and is complementary to target DNA sequence, which is programmable to target different DNA sites. The tracrRNA is a constant part of sgRNA, forms several stem-loops and is required for Cas9 binding, crRNA processing and Cas9-mediated target cleavage (Jiang and Doudna 2017). The ease of sgRNA design by simply changing the gRNA spacer sequence drives the use of CRISPR/Cas9 technology for performing different genetic perturbations and thus resulting in variable phenotypes.

The design of CRISPR/Cas9 gene-editing experiment begins with the selection of gRNA spacer sequence. A critical step of CRISPR/Cas9-sgRNA system is the optimum design of sgRNA. Although it is simple to find target sites in the genome by scanning PAM sequence sites, several challenges exist in the context of efficacy and specificity that in turn determines the precision of gene editing. Here, efficacy is defined as the likelihood that the particular sgRNA facilitates precise cutting, and specificity is defined as the likelihood that sgRNA binds to off-target sites. Cas9 exhibits variable efficiency at different target sites and accepts few base mismatches and DNA/RNA bulges (Lee et al. 2016). However, a careful study on the efficacy and specificity is needed to ensure the success of CRISPR/Cas9 experiment. Therefore, to assist researchers in performing CRISPR/Cas9 experiments, there is a need to develop prediction models/algorithms and computational tools to maximise on-target activity and minimise potential off-target effects (Doench et al. 2016).

Next-generation sequencing technology and bioinformatics have resulted in the annotation of genetic elements in the genome of different plant species. Annotated genome databases are being developed for different crops. Researchers can leverage this genome information for CRISPR studies in order to understand the factors contributing to on-target and off-target activity of CRISPR-Cas9 system. Models have been built that resulted in the development of computational tools for sgRNA design that assist researchers in the selection of best target sites. In CRISPR-mediated gene knock-out/in studies, the use of these tools has been well demonstrated (Parkhi et al. 2018; Jacob and Gregory 2016). Computational tools have been developed for gene regulation, genome-wide custom sgRNA libraries construction and to facilitate sgRNA design for custom gene sets in a single step. Further for detecting CRISPR induced mutations and mutation patterns, NGS platforms are being used to evaluate genome editing results. Bioinformatics tools have been developed to support researchers in NGS data analysis and result interpretation, primarily for identifying CRISPR introduced mutations and for large scale screening.

In this chapter, we have reviewed the on-target efficiency prediction models and off-target prediction algorithms used in different sgRNA design tools. We have compiled the information on the bioinformatics tools that are available currently for different CRISPR/Cas9 applications in the context of plant species.

2.2 On-Target Efficiency Prediction Models—An Overview

CRISPR/Cas9 system displays a high degree of variability in their cleavage activity. This shows that sgRNAs differ in their efficiency in Cas9 targeting (Doench et al. 2014; Lee et al. 2016; Wilson et al. 2018). Studies have reported that nucleotide composition of gRNA, position-specific nucleotides like the preference for a G immediately upstream of PAM, GC content, gRNA melting temperature, poly-T sequences, the thermodynamic stability of sgRNA, target site location relative to the Transcription Start Site (TSS) and target sites definition (only 20 bp target sequence and/including PAM/flanking sequence), have an influence on CRISPR/Cas9 activity (Doench et al. 2014; Graham and Root 2015; Wilson et al. 2018). Using experimental data as a training set along with different algorithms/machine learning methods, efficacy prediction models have been developed incorporating these features to predict sgRNA target efficacy (Cui et al. 2018). Rule Set 1 and 2 (Doench et al. 2014, 2016), SVM (Support Vector Machine) (Labaj et al. 2017), SVM Classifier model (Chari et al. 2015) and Elastic Net (Zou and Hastie 2005) are the most common models used in the sgRNA design tools (Table 2.1).

Predictive models have evolved from research that has analysed sgRNA editing efficiency in mammalian cells and a few of these models are being implemented in plants for evaluating sgRNA activity (Liang et al. 2016). However, it is still unclear whether the sequence features and design rules set up for optimising sgRNA in mammalian cells are reproducible across different sgRNA libraries or organisms (Graham and Root 2015; Chuai et al. 2016). A study by Liang et al. (2016) conducted on validated sgRNAs in plants had shown that unlike in animals, the nucleotide preferences were not found in plant sgRNAs implying a difference between plant and animal sgRNAs. They constituted a set of criteria for selection of efficient sgRNA in plants. Setting up a design rule for the selection of efficient sgRNA has been challenging in plants due to time and efforts required to generate edited plants.

Regarding model prediction efficiency, it has been observed that models were not consistently accurate across independent datasets and no single feature governs sgRNA activity. It has been observed that the model prediction efficiency is strongly influenced by training set experimental conditions and the dataset used, gRNA transcription method and the way CRISPR/Cas9 activity is measured (Wilson et al. 2018). Currently, no single model is accounting for gRNA transcription method. Comparison of learning models on specific datasets has been evaluated but their general performance across different datasets is not known (Chaui et al. 2016). In tools like CRISPOR (Haeussler et al. 2016), E-CRISP (Heigwer et al. 2014), CHOPCHOP (Montague et al. 2014) and CLD (Heigwer et al. 2016), multiple predictive models have been integrated to enhance activity prediction. Additionally, epigenetic features, evolutionary conservation and disorder feature of the target protein sequence (Chen et al. 2017) have been found to influence CRISPR/Cas9 activity. Currently, in silico on target tool like Deep*CRISPR* (Chuai et al. 2018) has incorporated these features. Computational research is ongoing to identify additional features that contribute to sgRNA efficacy. A general view has been that by incorporating a combination of features there is an improvement in the accuracy of predictive models (Chuai et al. 2016; Wilson et al. 2018).

2.3 Off-Target Prediction Algorithms—An Overview

The existing sgRNA design tools have used alignment tools such as BWA ((Li and Durbin 2009) in CRISPOR (Haeussler et al. 2016)), Bowtie ((Langmead et al. 2009) in CCTop (Stemmer et al. 2017)) and Penalty Matrix ((Hsu et al. 2013) in CHOPCHOP (Montague et al. 2014)) to align short target sequence against the reference genome. For identifying potential off-target sites, the sgRNA design tool's search for mismatch counts and positions falling in seed regions and accordingly assigns a score to off-target sites. Tools have an option to select different base mismatch tolerance, hence differ in off-target prediction results (Table 2.1). Studies comparing experimentally validated off-targets and sites predicted by alignment tools have come

Table 2.1 sgRNA design tools in plants

Tool name	Website	Maximum mismatches allowed	PAM sequence	Input	gRNA annotation	On-target prediction (efficiency)	Off-target scoring (specificity)	Description
Synthego design tool	https://design.synthego.com/#/	3	NGG	DNA sequence	Yes	Yes	Yes	Design top-scoring guide RNAs for gene knockout
Benchling CRISPR gRNA design	https://benchling.com/crispr	4	User customizable	Gene ID/Genome coordinates	Yes	Yes	Yes	Score gRNAs by specificity and efficiency and annotate gRNA
Breaking-Cas	http://bioinfogp.cnb.csic.es/tools/breakingcas/	4	User customizable	DNA sequence (<20,000 bases)	Yes	No	Yes	Supports all eukaryotic genomes available in the ENSEMBL database. Uses nuclease-specific scoring systems for assessing off-targets. Built-in mini genome browser for inspecting predicted targets and off-targets sites
Cas-OFFinder	http://www.rgenome.net/cas-offinder/	0–9	20 PAMs (NGG, NRG, NNAGAAW, …)	crRNA sequence (<1000 bases)	No	No (Alignment-based)	Yes, Cas-OFFinder (Bae et al. 2014)	Fast and highly versatile off-target searching tools, option for specifying mismatch number and RNA/DNA bulge size

(continued)

Table 2.1 (continued)

Tool name	Website	Maximum mismatches allowed	PAM sequence	Input	gRNA annotation	On-target prediction (efficiency)	Off-target scoring (specificity)	Description
CRISPOR	http://crispor.org/	4	19 PAMs (NGG, NGA, NGCG, NNAGAA, NNNNGMTT, NNNNACA, …)	DNA sequence/Genome coordinates	Yes	Yes (Learning-based): Rule Set2 (Doench et al. 2016), SVM (Labaj et al. 2017), Elastic-Net (Zou and Hastie 2005), SVM classifier model (Chari et al. 2015; Doench et al. 2014; Xu et al. 2015; Housden et al. 2015)	Yes, CFD score (Doench et al. 2016; Hsu et al. 2013)	The very interactive web tool, currently assists for >300 genomes. Predicts gRNA sequence (link to the genome browser), their efficiency, off-target and micro-homology scores. PCR primer design option for downstream analysis
CCTop	https://crispr.cos.uni-heidelberg.de/	5 (0–5)	11 PAMs (NGG, NNGRRT, NAAAAC, …)	DNA sequence (max. 500 bases), single or batch mode	Yes	Yes (Learning-based): Elastic-Net (Zou and Hastie 2005; Chari et al. 2015), CRISPRater (Labuhn et al. 2018; Doench et al. 2016; Xu et al. 2015)	Yes, (Stemmer et al. 2017)	Friendly interface ranks all possible sgRNA target sites based on off-target quality. Provides comprehensive information on each potential off-target site

(continued)

Table 2.1 (continued)

Tool name	Website	Maximum mismatches allowed	PAM sequence	Input	gRNA annotation	On-target prediction (efficiency)	Off-target scoring (specificity)	Description
CHOPCHOP	https://chopchop.rc.fas.harvard.edu/	0, 2	7 PAMs (NGG, NAG, NGA, NRG, NNNNGANN, …), User customizable	RefSeq/ENSEMBL/Gene ID/Genome coordinates/DNA sequence	No	Yes (Learning-based): Rule Set2 (Doench et al. 2016), SVM (Labaj et al. 2017), Elastic-Net (Zou and Hastie 2005; Chari et al. 2015; Xu et al. 2015; Moreno-Mateos et al. 2015)	Yes, Bowtie (Langmead et al. 2009; Cong et al. 2013; Hsu et al. 2013)	A widely used tool for CRISPR-Cas9/Cas9 nickase/Cpf1 and TALEN-based genome editing. The program ranks potential sgRNAs based on a number of off-targets and mismatched sites. Interactive visualization and feature of primer design
CHOPCHOP v2	http://chopchop.cbu.uib.no/	3 (0–3)	User customizable	Same as of CHOPCHOP	Yes	Yes (Learning-based: same as of CHOPCHOP)	Yes	Added sgRNA design also for CRISPR-Cpf1 and Cas9 nickase, an additional scoring system like self-complementarity scoring for sgRNAs to enhance efficiency and specificity. Currently, the tool supports 32 organisms and is open to add new genomes

(continued)

Table 2.1 (continued)

Tool name	Website	Maximum mismatches allowed	PAM sequence	Input	gRNA annotation	On-target prediction (efficiency)	Off-target scoring (specificity)	Description
CRISPR Configurator and Specificity Tool	https://dharmacon.horizondiscovery.com/resources/tools-and-calculators/	8 (gaps or mismatches)	NGG and NAG	DNA sequence/Genome coordinates/Acc. no.	Yes	No (Alignment-based)	No	Identify putative off-target alignments (even with a gap) for input genomic DNA sequences
CRISPRdirect	http://crispr.dbcls.jp/	Any number	NNN, User customizable	DNA sequence (≤10kbp)/Genome coordinates/Acc. no.	Yes	No (Alignment-based)	Yes, (Naito et al. 2015)	Supports *Arabidopsis*, rice and sorghum. The output window reports target position, sequence and the number of target sites in the genome with 8 or 12 bp matching seed sequences
CRISPR gRNA Design Tool	https://www.dna20.com/eCommerce/cas9/input	0-10	NGG, NAG	DNA sequence/Gene locus/Genome coordinates	Yes	No (Alignment-based)	Yes	Support sgRNA design for *Arabidopsis*

(continued)

Table 2.1 (continued)

Tool name	Website	Maximum mismatches allowed	PAM sequence	Input	gRNA annotation	On-target prediction (efficiency)	Off-target scoring (specificity)	Description
CRISPRseek	http://www.bioconductor.org/packages/release/bioc/html/CRISPRseek.html	Any number	User customizable	DNA sequence	Yes	Yes (Alignment-based) Learning—(Doench et al. 2014)	Yes, CFD score (Doench et al. 2016; Hsu et al. 2013)	Highly flexible, open-source Bioconductor package, runs on R. Includes functionalities for score target sites in two related input sequences
GT-Scan (Genome Target Scan)	https://gt-scan.csiro.au/	3 (0-3)	User customizable	DNA Sequence	No	Yes	Yes, (Hsu et. 2013)	Supports barley, maize, pepper, rice, soybean, tomato, *Arabidopsis*, foxtail millet and *Physcomitrella Patens*. Rank sgRNA targets by their uniqueness and interactive output window
CRISPR-P	http://crispr.hzau.edu.cn/cgi-bin/CRISPR/CRISPR	System defined	NGG, NAG	DNA sequence/Gene locus/Genome coordinates	Yes	Yes (Learning-based): Elastic-Net (Zou and Hastie 2005), SVM classifier model (Chari et al. 2015; Lei et al. 2014)	Yes, (Hsu et al. 2013)	Supports > 20 plant species, predict the off-target score and visual interface for sgRNA off-targets and restriction enzyme analysis

(continued)

Table 2.1 (continued)

Tool name	Website	Maximum mismatches allowed	PAM sequence	Input	gRNA annotation	On-target prediction (efficiency)	Off-target scoring (specificity)	Description
CRISPR-P 2.0	http://crispr.hzau.edu.cn/cgi-bin/CRISPR2/CRISPR	System defined	14 PAMs (NGG, NNAGAAW, NNNNGMTT, TTTN, …)	Gene locus/Genome coordinates	Yes	Yes (Learning-based): Elastic-Net (Zou and Hastie 2005), SVM classifier model (Chari et al. 2015; Doench et al. 2016)	Yes	Supports 49 plant genomes and predict sgRNA on-and off-target activities. Provide information on micro-homology score, GC content and sgRNA secondary structure
Crispr-Plant	https://www.genome.arizona.edu/crispr/	System defined	NGG	Gene locus/Genome coordinates	Yes	Yes (Learning-based): Rule set2 (Doench et al. 2016)	Yes	Supports seven model plant species (*A. thaliana, M. truncatula, G. max, S. lycopersicum, B. distachyon, O. sativa, S. bicolor and Z. mays*). gRNA spacer sequences classified into various classes like Class0, class1, class2 based on the number of mismatches
E-CRISP	http://www.e-crisp.org/E-CRISP/	System defined	NGG, User Customizable	Gene ID/Gene symbol/DNA Sequence	Yes	Yes (Learning-based): SVM classifier model (Chari et. 2015; Doench et al. 2014; Xu et al. 2015), E-score (Heigwer et al. 2014)	S-score (Heigwer et al. 2014)	Supports rice, wheat, barley, maize and grapevines. Option for single gene/batch design through CLD (Heigwer et al. 2016). Output display statistics on sgRNA targets and target gene

(continued)

Table 2.1 (continued)

Tool name	Website	Maximum mismatches allowed	PAM sequence	Input	gRNA annotation	On-target prediction (efficiency)	Off-target scoring (specificity)	Description
CRISPR-GE	http://skl.scau.edu.cn/	System defined	NGG, TTN, TTTN, User Customizable	DNA sequence/Gene symbol/Gene ID	Yes	Yes	Yes	Integrated tool kit for CRISPR-Cas9/cpf1-based genome editing, experimental analysis to mutation detection, includes a set of tools for target design (Target Design), off-target sites (off-target) and primer design
CRISPR MultiTargeter	http://www.multicrispr.net/	System defined	NGG, User Customizable	DNA sequence/RefSeq/ENSEMBL/Gene ID	No	Yes (Learning-based): (Doench et al. 2014)	Yes	Find common and unique sgRNA targets in a set of similar sequences. Results redirect to GT-Scan (Adian and Bailey 2014) or Cas-OFFinder (Bae et al. 2014) for off-target analysis
CGAT (CRISPR Genome Analysis Tool)	http://cbc.gdcb.iastate.edu/cgat/	System defined	NGG	Gene ID/DNA sequence	No	No (Alignment-Based)	Yes	For barley, maize, peanut, rice and soybean, predicts off-targets scores and ranks sgRNAs. All of these functionalities contain within a single pipeline

(continued)

Table 2.1 (continued)

Tool name	Website	Maximum mismatches allowed	PAM sequence	Input	gRNA annotation	On-target prediction (efficiency)	Off-target scoring (specificity)	Description
CT-Finder	http://bioinfolab.miamioh.edu/ct-finder	System defined	NGG	DNA sequence	Yes	No	No	For rice, maize and *Arabidopsis*, design sgRNA also for Cas9 nickase and RNA-guided *FokI* nucleases (RFNs) and provides on and off-target sites graphical visualisation

across false-positive (i.e. off-targets are actually cleaved by CRISPR/Cas9 or overestimation) and false-negative (i.e. off-targets significantly differ from target site or underestimation) target site predictions (Wilson et al. 2018). To balance the false positive and false negative predictions and to filter out false positives, scoring algorithms are being developed. The two scoring methods developed are the MIT-Broad (Hsu et al. 2013) and CFD Score (Doench et al. 2016). In the recently developed off-target methods by Elevation (Listgarten et al. 2018) and CRISTA (Abadi et al. 2017), additional features such as gRNA secondary structure, off-target site overlap with DNase-sensitive regions and type of mismatch (wobble vs. DNA/RNA bulge) have been included in the model, and better prediction of off-target sites was observed. There is also evidence of chromatin feature influencing Cas9 binding, a feature for the first time included in CROP-IT pipeline (Singh et al. 2015). As a single tool is not optimal for all conditions, Zhang et al. (2018) proposed synergising multiple tools to enhance off-target predictive capabilities. Incorporation of these factors into the models could help in the further refinement in the current in silico prediction methods (Wilson et al. 2018).

2.4 Bioinformatics Tools for CRISPR/Cas9 Applications in Plants

For conducting CRISPR experiments and postediting analysis, different application-specific bioinformatics tools have been developed and have been summarised below. Schematic overview of CRISPR bioinformatics tools classified as per the applications is shown in Fig. 2.1.

2.4.1 CRISPR-Mediated Knock-Out/in

Currently, over 60 sgRNA design tools have been developed for the gene-by-gene design to carry out CRISPR knock-out/in studies. These tools provide a degree of flexibility to users with respect to input sequence, design specifications and parameter choices (option to select PAM sequence, base mismatch number, gRNA sequence length), graphical visualisation and downstream analysis. In the context of plants, Table 2.1 lists the tools, along with a description of the functionality of each tool that is being commonly used for designing sgRNA. CRISPR-Plant (Xie et al. 2014), CRISPR-P (Lei et al. 2014) and CRISPR-P 2.0 (Liu et al. 2017) were specifically developed for plants for the design of sgRNA. Tools such as the GT-Scan (Adian and Bailey 2014), E-CRISP (Heigwer et al. 2014) are restricted to specific genomes (Table 2.1) while other tools like CRISPRseek (Zhu et al. 2014) (Table 2.1) are generic, where a user-specified sequence or genome can be given as input. Generic tools like Crisflash (Jacquin et al. 2019), DESKGEN (https://www.deskgen.com/lan

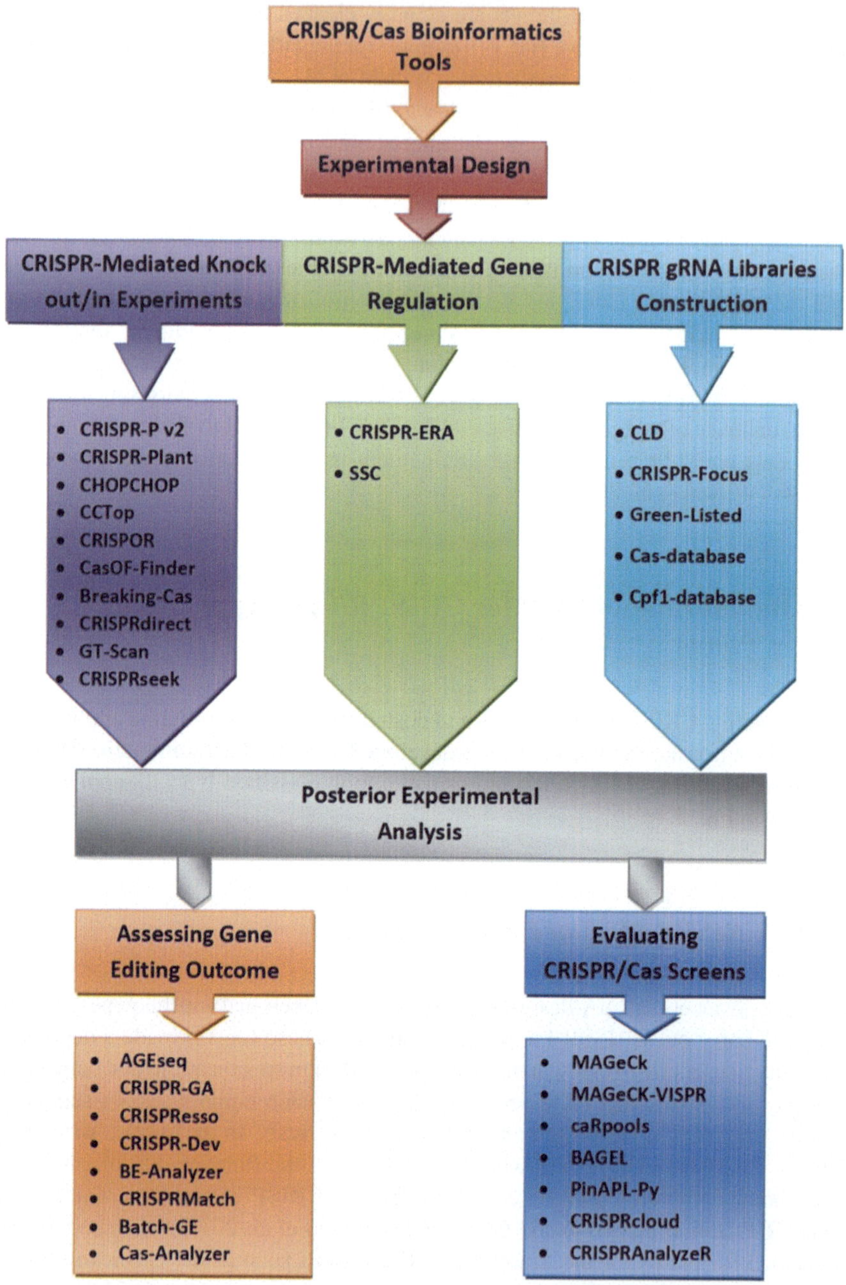

Fig. 2.1 Schematic of different application-specific CRISPR bioinformatics tools

ding/), COD (http://cas9.wicp.net/), Genedata Selector (https://www.genedata.com/) and CRISPR LifePipe (https://crispr.lifeandsoft.com/) allow the design of sgRNA for non-model organisms or custom genome sequences.

Newer sgRNA tools offer features to accommodate variants of Cas9 that utilise alternative PAMs, with an option to select either predefined (CRISPOR (Haeussler et al. 2016), Cas-OFFinder (Bae et al. 2014)) or user-defined PAMs (Breaking-Cas (Oliveros et al. 2016)). Tools such as the CRISPR MultiTargeter (Prykhozhij et al. 2015) and CRISPys (Hyams et al. 2017) have been developed to identify common and unique target sites from a set of similar DNA sequences. All these tools (Table 2.1) differ in the output display format. While some tools provide the complete list of PAM sites (CRISPRdirect (Naito et al. 2015)), other tools have an interactive graphical interface, linking to genome browser/target gene that enable users to select optimal sites like in CRISPOR (Haeussler et al. 2016). Tools like CCTop (Stemmer et al. 2017), CRISPR-P (Lei et al. 2014), CRISPR-P 2.0 (Liu et al. 2017) follow target scoring metrics (on-target sgRNA efficiency prediction and/or off-target prediction) to rank all possible candidate sgRNAs. Cas-OFFinder (Bae et al. 2014), E-CRISP (Heigwer et al. 2014), CHOPCHOP (Montague et al. 2014), CRISPR-P (Lei et al. 2014), CRISPRdirect (Naito et al. 2015), GT-Scan (Adian and Bailey 2014), CRISPRseek (Zhu et al. 2014) have the strength to predict potential off-target score or binding sites, that is the specificity of sgRNA and rank candidate sgRNAs based on specificity (Table 2.1). Cas-OFFinder (Bae et al. 2014) has been designed specifically for searching potential off-target sites from a given PAM containing the sequence. Based on the method, tools have been classified into two types, (i) Alignment-based: where the sgRNAs are aligned by locating the PAM site and (ii) Learning-based: where sgRNAs are predicted and scored, based on the predictive models and different features affecting the efficacy also been considered (Table 2.1).

2.4.2 CRISPR-Mediated Gene Regulation

Advances in CRISPR engineering tools have led to the development of inactivated catalytically dead Cas9 (dCas9). When dCas9 is fused with activator or repressor domains, control of gene expression is achieved (Qi et al. 2013). A set of rules to predict effective and specific sgRNAs for CRISPR activation and repression have been identified which have been used to generate genome-scale CRISPRi and CRISPRa libraries targeting human and mouse genomes (Gilbert et al. 2013; Horlbeck et al. 2016). However till date, a limited number of sgRNA design tools have been developed for gene regulation. CRISPR-ERA (Liu et al. 2015) is the first tool that attempted to design sgRNAs for gene repression or activation. This tool provides sgRNA design for nine model organisms (prokaryotes, animals and humans), ranks sgRNAs by the sum of E- and S-scores, and target sites can be visualised through genome browser. Sequence Scan for CRISPR (SSC) is another sgRNA tool wherein Xu et al. 2015 used CRISPRi/a screening data and developed a model for predicting sgRNA efficiency for CRISPRi/a experiments. In the context of sequence features, they found a substantial difference in sequence preference for CRISPRi/a from that of CRISPR/Cas9 knock-out attributing to the distinct mechanism used by CRISPRi/a system to perturb gene function. For example, sgRNA with a spacer length of 19 nt were found to be better than 20 nt spacers for CRISPRi/a experiments. However, information regarding the prediction of sgRNA efficiency in CRISPRi/a has been derived from studies in bacterial, yeast, mouse and human cells (Qi et al. 2013; Jensen 2018). In plants, CRISPRi/a system is still an upcoming platform for use as a functional genomics tool for gene regulation. CRISPR/dcas9 system has been established in a few plant species such as *Nicotiana benthamiana* (Piatek et al. 2015) and *A. thaliana* (Park et al. 2017b). Sufficient plant-specific CRISPRi/a data is required to validate the existing/set up the sgRNA design rules.

2.4.3 CRISPR sgRNA Libraries Construction

CRISPR/Cas9 technology has promising applications for high-throughput functional genetic screening. Studies on CRISPR-screen used general criteria/sgRNA design tools (capacity of single gene/nominal batch design) to construct genome-wide sgRNA libraries (Meng et al. 2017). However, with a growing need for sgRNA libraries for custom gene sets and for optimal library design, specific tools are being designed for an end-to-end design of custom sgRNA libraries targeting specific genomes. While designing algorithms for CRISPR screens, parameters that determine on-target efficiency and off-target specificity have been considered to increase the overall efficiency of sgRNA libraries. CRISPR library designer (CLD) for sgRNA libraries, developed by Heigwer et al. 2016 is first such tool, to target a few hundred genes to genome-scale and is suitable for all annotated organisms. However, CLD works best on well-annotated genomes and is sensitive to off-target sites prediction (Heigwer et al. 2016). CRISPR-Focus (Cao et al. 2017) designed for human and mouse genomes, can target up to one thousand genes with minimum user input. Green-Listed (Panda et al. 2017) is designed for human and mouse, facilitates the design of custom CRISPR screens against a selected list of genes. Cas-database (Park et al. 2016) allows users to select optimal target sequences at once from thousands of genes for genome-wide sgRNA library construction. This tool can be applied to 12 different organisms that include five plants species, *Arabidopsis*, banana, tomato, grapevine and soybean. Park and Baes (2018) developed Cpf1-Database, a genome-wide sgRNA library design tool for LbCpf1 and AsCpf1 endonucleases, a type V CRISPR-Cas9 system and this database supports *Arabidopsis*, grapevine, soybean, banana and tomato genomes. CRISPR-Local (Sun et al. 2018) facilitates the design of genome-wide sgRNAs for non-reference plant genome.

2.4.4 *Posterior Experimental Analysis*

2.4.4.1 Assessing Gene Editing Outcome

Amplicon sequencing (commonly performed for gene editing analysis) using next-generation platforms is being used for screening of CRISPR/Cas9 induced clones, attributing to cost-effectiveness, deeper coverage and high throughput in terms of sample numbers and multiplexed targets (Bell et al. 2014). Web-based tools have been developed to process the high-throughput genome editing NGS data. These tools calculate mutation frequency, evaluate editing efficiency, accuracy and provide comprehensive visualisation of results. Guell et al. 2014 provided a platform, CRISPR-GA, where user can upload amplified reads and original sequence (one-by-one sample analysis) and upon submission the user gets an estimate on insertions, deletions and homologous recombination (HR) and corresponding frequencies at the desired targeting sites. AGEseq (Xue and Tsai 2015) is a standalone program that allows analysis against multiple reference sequences at a time, a feature not available in CRISPR-GA tool. CRISPResso (Pinello et al. 2016) offers features that allow batch sample analysis via a command-line interface, compatible with other bioinformatics pipelines, improve quantification accuracy and output visualisation. Batch-GE (Boel et al. 2016) offers a batch analysis of multiple sample runs on Linux-based server or in a standalone mode. However, all of these tools take a very long time to upload large read files. To address this, Cas-Analyzer (Park et al. 2017a) allows data uploading on the client-side, thus reducing time-consuming step of loading to the server. CRISPR-Dav (Wang et al. 2017) runs in Perl and R and allows analyses of multiple samples in parallel or serially, features detection of large INDELS and improved visualisation. CRISPRMatch (You et al. 2018) is a standalone program and recently Hwang et al. 2018 designed a web-based tool, BE-Analyzer to analyse NGS data for CRISPR-base editor experimental results.

2.4.4.2 Evaluating CRISPR/Cas9 Screens

To analyse and interpret pooled library NGS data generated through CRISPR/Cas9 screens, bioinformatics tools have been developed to support scientists in identifying and selecting edited candidate genes. These tools aim to provide complete/end-to-end analysis solution, beginning with analysing raw sequence reads and ending with a ranking of edited candidate genes. MAGeCk (Li et al. 2014) is the first command-line software that requires knowledge of read sequence composition thereby limiting accessibility to a larger research community. EdgeR (Dai et al. 2014) is a bioconductor package, BAGEL (Hart and Moffat 2016) is python-based standalone tool, and caRpools (Winter et al. 2016) runs on R platform that requires R proficiency for analysing pooled CRISPR screens data. MAGeCK-VISPR (Li et al. 2015) presents a quality control and visualisation workflow for CRISPR screens. As compared to tools that require additional tools to complete the analysis, CRISPRAnalyzeR

(Winter et al. 2017) is the first end to end analysis platform developed for pooled CRISPR/Cas9 screens. The power of this tool is that it uses eight different algorithms simultaneously to identify edited candidate genes/CRISPRcloud (Jeong et al. 2017) and PinAPL-Py (Spahn et al. 2017) are user-friendly web-interface tools that allow analysis on paired-end data. PinAPL-Py is an improvement over CRISPRcloud in terms of higher flexibility (automatic sgRNA extraction feature), functionality in read alignment, sgRNA and gene ranking.

2.5 Concluding Remarks and Future Perspective

Computational tools have been developed to ease the experimental design in applications of CRISPR/Cas9 technology as well as for analysing gene-edited high-throughput data. For enhancing the efficiency of CRISPR-mediated experiments either for gene editing/gene regulation/functional screening, it is necessary to use computational tools for the prediction of sgRNA efficacy and specificity. Through bioinformatics and computational techniques, models and algorithms have been developed to facilitate optimal sgRNA design. Several sgRNA design tools for plant genomes including non-reference plant species are available in the public domain to assist the researcher. Some tools have the strength in predicting off-target sites more accurately while others have the potential to predict both on-target activity and off-target specificity. Tools have been developed for end-to-end design of custom sgRNA libraries targeting different genomes to facilitate optimal library design. However, specific models for predicting sgRNA efficiency in plants is still lacking and further studies are required to understand the efficiency of sgRNAs in plants. In addition, a better understanding of factors that contribute to sgRNA on-target and off-target effects is required in order to increase the accuracy of predictive models.

References

Abadi S, Yan WX et al (2017) A machine learning approach for predicting CRISPR-Cas9 cleavage efficiencies and patterns underlying its mechanism of action. PLoS Comput Biol 13:e1005807

Adian OB, Bailey TL (2014) GT-Scan: identifying unique genomic targets. Bioinformatics 30:2673–2675

Bae S, Park J, Kim JS (2014) Cas-OFFinder: a fast and versatile algorithm that searches for potential off-target sites of Cas9 RNA-guided endonucleases. Bioinformatics 30:1473

Belhaj K, Chaparro-Garcia A et al (2013) Plant genome editing made easy: targeted 273 mutagenesis in model and crop plants using the CRISPR/Cas system. Plant Methods 9:39

Bell CC, Graham WM et al (2014) A high-throughput screening strategy for detecting CRISPR-Cas9 induced mutations using next-generation sequencing. BMC Genom 15:1002

Boel A, Steyaert W, Rocker ND et al (2016) BATCH-GE: batch analysis of next-generation sequencing data for genome editing assessment. Sci Rep 6:30330

Bortesi L, Fischer R (2015) The CRISPR/Cas9 system for plant genome editing and beyond. Biotechnol Adv 33:41–52

Cao Q, Ma J, Chen CH et al (2017) CRISPR-FOCUS: a web server for designing focused CRISPR screening experiments. PLoS ONE 12:e0184281

Chari R, Mali P et al (2015) Unraveling CRISPR-Cas9 genome engineering parameters via a library-on-library approach. Nat Methods 12:823

Chen L, Wang S, Zhang YH et al (2017) Identify key sequence features to improve CRISPR sgRNA efficacy. IEEE Access 5:26582–26590

Chuai G, Ma H, Yan J, Chen M et al (2018) Deep*CRISPR*: optimized CRISPR guide RNA design by deep learning. Genome Biol 19:80

Chuai GH, Wang QL, Qi L (2016) *In silico* meets *in vivo*: towards computational CRISPR-based sgRNA design. Trends Biotechnol 35:12

Collonnier C, Debast AG et al (2017) Towards mastering CRISPR-induced gene knock-in in plants: survey of key features and focus on the model *Physcomitrella patens*. Methods 121–122:103–117

Cong L, Ran FA, Cox D et al (2013) Multiplex genome engineering using CRISPR/Cas systems. Science 339:819–823

Cui Y, Xu J, Cheng M et al (2018) Review of CRISPR/Cas9 sgRNA design tools. Interdiscip Sci Comput Life Sci 10:455–465

Dai Z, Sheridan JM, Gearing LJ et al (2014) edgeR: a versatile tool for the analysis of shRNA-seq and CRISPR-Cas9 genetic screens. F1000Research 3:95

Doench JG, Fusi N, Sullender M et al (2016) Optimized sgRNA design to maximize activity and minimize off-target effects of CRISPR-Cas9. Nat Biotechnol 34:184–191

Doench JG, Hartenian E, Graham DB et al (2014) Rational design of highly active sgRNAs for CRISPR-Cas9-mediated gene inactivation. Nat Biotechnol 32:1262–1267

Gilbert LA, Horlbeck MA, Adamson B et al (2013) CRISPR-mediated modular RNA-guided regulation of transcription in eukaryotes. Cell 154:442–451

Graham DB, Root DE (2015) Resources for the design of CRISPR gene editing experiments. Genome Biol 16:260

Guell M, Yang L, Church GM (2014) Genome editing assessment using CRISPR Genome Analyzer (CRISPR-GA). Bioinformatics 30:2968–2970

Haeussler M, Schonig K, Eckert H et al (2016) Evaluation of off-target and on-target scoring algorithms and integration into the guide RNA selection tool CRISPOR. Genome Biol 17:148

Hart T, Moffat J (2016) BAGEL: a computational framework for identifying essential genes from pooled library screens. BMC Bioinform 17:164

Hashimoto R, Ueta R, Abe C et al (2018) Efficient multiplex genome editing induces precise, and self-ligated type mutations in tomato plants. Front Plant Sci 9:916

Heigwer F, Kerr G, Boutros M (2014) E-CRISP: fast CRISPR target site identification. Nat Methods 11:122–123

Heigwer F, Zhan T, Breinig M et al (2016) CRISPR library designer (CLD): software for multispecies design of single guide RNA libraries. Genome Biol 17:55. https://doi.org/10.1186/s13059-016-0915-2

Horlbeck MA, Gilbert LA, Villalta JE et al (2016) Compact and highly active next-generation libraries for CRISPR-mediated gene repression and activation. eLife. https://doi.org/10.7554/elife.19760

Housden BE, Valvezan AJ, Kelley C et al (2015) Identification of potential drug targets for tuberous sclerosis complex by synthetic screens combining CRISPR-based knockouts with RNAi. Sci Signal 8(393):rs9. https://doi.org/10.1126/scisignal.aab3729

Hsu PD, Scott DA, Weinstein JA et al (2013) DNA targeting specificity of RNA-guided Cas9 nucleases. Nat Biotechnol 31:827–832

Hussain B, Lucas SJ, Budak H (2018) CRISPR/Cas9 in plants: at play in the genome and at work for crop improvement. Brief Funct Genomics 17:319–328

Hwang GH, Park J, Lim K et al (2018) Web-based design and analysis tools for CRISPR base editing. BMC Bioinform 19:542

Hyams G, Abadi Avni A et al (2017) CRISPys: Optimal sgRNA design for editing multiple members of a gene family using the CRISPR system. J Mol Biol 430(15):2184–2195. https://doi.org/10.1101/221341

Jacob TB, Gregory BM (2016) High-throughput CRISPR vector construction and characterization of DNA modifications by generation of tomato hairy roots. J Vis Exp 110:e53843

Jacobs TB, Zhang N et al (2017) Generation of a collection of mutant tomato lines using pooled CRISPR libraries. Plant Physiol 174:2023–2037

Jacquin ALS, Odom DT, Lukk M (2019) Crisflash: open-source software to generate CRISPR guide RNAs against genomes annotated with individual variation. Bioinformatics. https://doi.org/10.1093/bioinformatics/btz019

Jensen MK (2018) Design principles for nuclease-deficient CRISPR-based transcriptional regulators. FEMS Yeast Res 18:foy039

Jeong HH, Kim SY, Rousseaux MWC et al (2017) CRISPRcloud: a secure cloud-based pipeline for CRISPR pooled screen deconvolution. Bioinformatics 33:2963–2965

Jiang F, Doudna JA (2017) CRISPR–Cas9 structures and mechanisms. Annu Rev Biophys 46:505–529

Jung C, Gossmann GC, Braatz J et al (2017) Recent developments in genome editing and applications in plant breeding. Plant Breed 137:1–9

Labaj W, Papiez A et al (2017) Comprehensive analysis of MILE gene expression data set advances discovery of leukaemia type and subtype biomarkers. Interdiscip Sci Comput Life Sci 9:24–35

Labuhn M, Adams FF, Ng M et al (2018) Refined sgRNA efficacy prediction improves large- and small-scale CRISPR-Cas9 applications. Nucleic Acids Res 46:1375–1385

Langmead B, Trapnell C et al (2009) Ultrafast and memory-efficient alignment of short DNA sequences to the human genome. Genome Biol 10:R25

Lee CM, Cradick T et al (2016) Nuclease target site selection for maximizing on-target activity and minimizing off-target effects in genome editing. Mol Ther 24:475–487

Lei Y, Lu L, Liu HY et al (2014) CRISPR-P: a web tool for synthetic single-guide RNA design of CRISPR-system in plants. Mol Plant 7:1494–1496

Li H, Durbin R (2009) Fast and accurate short read alignment with Burrows-Wheeler transform. Bioinformatics 25:1754–1760

Li JF, Norville JE, Aach J et al (2013) Multiplex and 304 homologous recombination-mediated genome editing in *Arabidopsis* and *Nicotiana benthamiana* using guide RNA and Cas9. Nat Biotechnol 31:688–691

Li W, Koster J, Xu H et al (2015) Quality control, modeling, and visualization of CRISPR screens with MAGeCK-VISPR. Genome Biol 16:281

Li W, Xu H, Xiao T et al (2014) MAGeCK enables robust identification of essential genes from genome-scale CRISPR/Cas9 knockout screens. Genome Biol 15:554

Liang G, Zhang H et al (2016) Selection of highly efficient sgRNAs for CRISPR/Cas9-based plant genome editing. Sci Rep 6:21451

Listgarten J, Weinstein M, Kleinstiver BP et al (2018) Prediction of off-target activities for the end-to-end design of CRISPR guide RNAs. Nat Biomed Eng 2:38–47

Liu H, Wei Z, Dominguez A et al (2015) CRISPR-ERA: a comprehensive design tool for CRISPR-mediated gene editing, repression and activation. Bioinformatics 31:3676–3678

Liu H, Ding Y, Zhou Y et al (2017) CRISPR-P 2.0: an improved CRISPR-Cas9 tool for genome editing in plants. Mol Plant 10:530–532

Lu Y, Ye X, Guo R et al (2017) Genome-wide targeted mutagenesis in rice using the CRISPR/Cas9 system. Mol Plant 10:1242–1245

Meng X, Yu H, Zhang Y et al (2017) Construction of a genome-wide mutant library in rice using CRISPR/Cas9. Mol Plant 10:1238–1241

Montague TG, Cruz JM et al (2014) CHOPCHOP: a CRISPR/Cas9 and TALEN web tool for genome editing. Nucleic Acids Res 42:401–407

Moreno-Mateos MA, Vejnar CE, Beaudoin JD et al (2015) CRISPRscan: designing highly efficient sgRNAs for CRISPR-Cas9 targeting *in vivo*. Nat Methods 12:982

Naito Y, Hino K et al (2015) CRISPRdirect: software for designing CRISPR/Cas guide RNA with reduced off-target sites. Bioinformatics 31:1120–1123

Oliveros JC, Monica F et al (2016) Breaking-Cas-interactive design of guide RNAs for CRISPR-Cas experiments for ENSEMBL genomes. Nucleic Acids Res 44(W1):267–271

Panda SK, Boddul SV, Jiménez-Andrade GY et al (2017) Green listed-a CRISPR screen tool. Bioinformatics 33:1099–1100

Park J, Bae S (2018) Cpf1-Database: web-based genome-wide guide RNA library design for gene knockout screens using CRISPR-Cpf1. Bioinformatics 34:1077–1079

Park J, Kim JS, Bae S (2016) Cas-Database: web-based genome-wide guide RNA library design for gene knockout screens using CRISPR-Cas9. Bioinformatics 32:2017–2023

Park J, Lim K et al (2017a) Cas-analyzer: an online tool for assessing genome editing results using NGS data. Bioinformatics 33:286–288

Park JJ, Dempewolf E et al (2017b) RNA-guided transcriptional activation via CRISPR/dCas9 mimics overexpression phenotypes in *Arabidopsis*. PLoS ONE 12:e0179410

Parkhi V, Bhattacharya A, Choudhary S et al (2018) Demonstration of CRISPR-*cas*9-mediated *pds* gene editing in a tomato hybrid parental line. Indian J Genet 78:132–137

Piatek A, Ali Z, Baazim H et al (2015) RNA-guided transcriptional regulation in planta via synthetic dCas9-based transcription factors. Plant Biotechnol 13:578–589

Pinello L, Canver MC, Hoban MD et al (2016) Analyzing CRISPR genome-editing experiments with CRISPResso. Nat Biotechnol 34:695–697

Prykhozhij SV, Rajan V et al (2015) CRISPR MultiTargeter: a web tool to find common and unique CRISPR single guide RNA targets in a set of similar sequences. PLoS ONE 10:e0119372

Qi LS, Larson MH, Gilbert LA et al (2013) Repurposing CRISPR as an RNA-guided platform for sequence-specific control of gene expression. Cell 152:1173–1183

Singh R, Kuscu C et al (2015) Cas9-chromatin binding information enables more accurate CRISPR off-target prediction. Nucleic Acids Res 43:e118

Spahn PN, Bath T, Weiss RJ et al (2017) PinAPL-Py: a comprehensive web application for the analysis of CRISPR/Cas9 screens. Sci Rep 7:15854. https://doi.org/10.1038/s41598-017-16193-9

Stemmer M, Thumberger T et al (2017) CCTop: an intuitive, flexible and reliable CRISPR/Cas9 target prediction tool. PLoS ONE 12:e0176619

Sun J, Liu H, Liu H et al (2018) CRISPR-local: a local single-guide RNA (sgRNA) design tool for non-reference plant genomes. Bioinformatics. https://doi.org/10.1093/bioinformatics/bty970

Wang M, Mao Y, Lu Y et al (2018) Multiplex gene editing in rice with simplified CRISPR-Cpf1 and CRISPR-Cas9 systems. J Integr Plant Biol 60:626–631

Wang X, Tilford C, Neuhaus I et al (2017) CRISPR-DAV: CRISPR NGS data analysis and visualization pipeline. Bioinformatics 33:3811–3812

Wilson LOW, Brien ARO, Bauer DC (2018) The current state and future of CRISPR-Cas9 gRNA design tools. Front Pharmacol 9:749

Winter J, Breinig M, Heigwer F et al (2016) caRpools: an R package for exploratory data analysis and documentation of pooled CRISPR/Cas9 screens. Bioinformatics 32:632–634

Winter J, Schwering M, Pelz O et al (2017) CRISPRAnalyzeR: interactive analysis, annotation and documentation of pooled CRISPR screens. bioRxiv. https://doi.org/10.1101/109967

Xie K, Zhang J, Yang Y (2014) Genome-wide prediction of highly specific guide RNA spacers for the CRISPR-Cas9 mediated genome editing in model plants and major crops. Mol Plant 7:923–926

Xu H, Xiao T, Chen CH et al (2015) Sequence determinants of improved CRISPR sgRNA design. Genome Res 25:1147–1157

Xue LJ, Tsai CJ (2015) AGEseq: analysis of genome editing by sequencing. Mol Plant 8:1428–1430

Yin K, Qiu JL (2019) Genome editing for plant disease resistance: applications and perspectives. Philos Trans R Soc B Biol Sci 374(1767). https://doi.org/10.1098/rstb.2018.0322

You Q, Zhong Z, Ren Q et al (2018) CRISPRMatch: an automatic calculation and visualization tool for high-throughput CRISPR genome-editing data analysis. Int J Biol Sci 14:858–862

Zhang S, Li X et al (2018) Synergizing CRISPR/Cas9 off-target predictions for ensemble insights and practical applications. Bioinformatics. https://doi.org/10.1093/bioinformatics/bty748

Zhu LJ, Holmes BR et al (2014) CRISPRseek: a bioconductor package to identify target-specific guide RNAs for CRISPR-Cas9 genome-editing systems. PLoS ONE 9:e108424

Zou H, Hastie T (2005) Regularization and variable selection via the elastic net. J R Stat Soc 67:301–320

Chapter 3
CRISPR and Food Security: Applications in Cereal Crops

Mayank Rai, P. Magudeeswari, and Wricha Tyagi

Abstract The world population is expected to increase to 10 billion by the year 2050. The outbreak of new pests, diseases and reduction in agricultural arable land leads to yield loss and food insecurity. Among several cultivated crops, cereals are the major source of food to the world population. Along with molecular breeding methods to increase yield, modern biotechnological tools need to be employed to meet out these future demands. CRISPR/Cas9 is an efficient precise gene-editing tool that creates double-stranded breaks by using site-specific nuclease enzyme called Cas9 nuclease. Till date, efforts have been made to modify genes targeting yield increase, grain and nutritional quality, resistance to biotic and abiotic stress, development of male sterile lines, increasing dormancy duration, etc., using CRISPR-based non-transgenic genome editing tool. The available data suggests that the application of CRISPR technology in a timely manner would complement and supplement conventional and molecular plant breeding approaches to ensure global food and nutritional security in future.

Keywords Food security · Gene editing · CRISPR/Cas9 · Cereals

3.1 Introduction

Today, the major challenge faced by the human race is to feed the growing population. The world population is increasing rapidly and is expected to reach 10 billion by 2050 (Clarke and Zhang 2013). The population growth observed is more in tropical countries than in temperate countries. According to FAOSTAT (2016), the world food production needs to be increased by around 60–100% to fulfil this demand. Among all countries, seven tropical countries are expected to show more than 50% growth in upcoming years. These countries are India, Nigeria, the Democratic Republic of the Congo, Ethiopia, United Republic of Tanzania, Indonesia and Uganda (Campos and Caligari 2017). With issues such as extreme weather, reduced arable land, disease

M. Rai (✉) · P. Magudeeswari · W. Tyagi
School of Crop Improvement, College of Post Graduate Studies in Agricultural Sciences, Central Agricultural University (Imphal), Umroi road, Umiam 793103, Meghalaya, India
e-mail: mayank.rai@gov.in; mrai.cau@gmail.com

A. Bhattacharya et al. (eds.), *CRISPR/Cas Genome Editing*, Concepts and Strategies in Plant Sciences, https://doi.org/10.1007/978-3-030-42022-2_3

and pest outbreak and other abiotic stresses, it is imperative to adopt more advanced and improved technologies which can help to address the food security issue faster. In the last few decades conventional, mutation breeding and molecular breeding has been used to increase productivity. In order to achieve enhanced genetic gains in terms of yield in the shortest possible period of time, plant breeders have focused on improving the selection efficiency in crop breeding programmes through high throughput marker-assisted selection, genomic selection and precise phenomics-based selection technologies. Plant breeders have always envisaged an ideal haplotype (an individual carrying all the favourable alleles for all the important traits in a crop), however, the probability of getting such an individual through "recombinational breeding" is remote because of undesirable linkages with many such favourable alleles.

Recently, several sequence-specific genome editing technologies have emerged as useful tools for crop improvement (Georges and Ray 2017). Compared to ZFNs (Zinc Finger Nucleases) and TALENs (Transcription Activator-Like Effector Nucleases), CRISPR (Clustered Regularly Interspaced Short Palindromic Repeats)/CRISPR Associated protein 9 (Cas9) is more accurate in precise genome editing. This CRISPR technology can modify the genomic sequences that are present upstream of Protospacer Adjacent Motif (PAM) sequence to achieve desired traits.

3.2 Construction of CRISPR/Cas System and Improvements Thereof

CRISPR locus consists of tandem direct repeat sequences and spacer between the repeat sequences called CRISPR RNA (crRNA) (Kim and Kim 2014). The crRNA combines with another small RNA molecule which is complementary to crRNA repeats, known as transactivating CRISPR RNA (tracrRNA), and helps in processing of crRNA. After processing, crRNA and tracrRNA complex activates and guides the CRISPR associated (Cas) nucleases (generally Cas9), to the target site. The targeting of Cas nuclease into the specific site by crRNA and tracrRNA, leads to cleavage of homologous double-stranded DNA sequences, termed as a protospacers (Barrangou et al. 2007). Downstream of the protospacers sequences, the conserved sequence usually 5′-NGG-3′ (N=A or T or G or C) or less frequently 5′-NAG-3′ termed as protospacer adjacent motif (PAM) is present, which is essential for Cas mediated cleavage of the target site (Hsu et al. 2013). For applications in crop improvement, the crRNA can be engineered to target any DNA sequences, by taking the spacer sequences from the target region preceding the PAM sequence. For the ease of transformation and to increase its efficiency, crRNA and tracrRNA can be fused artificially to form single guide RNA (Mali et al. 2013; Qi et al. 2013). The CRISPR/Cas9 (CRISPR-associated) is the most widely used RNA-guided editing system. Researchers have successfully used CRISPR/Cas system to edit target gene in several crops (Table 3.1). But still, some factors which hinder the editing efficiency

Table 3.1 Summary of recent applications of CRISPR/Cas in cereal crops

Crop	Target gene	Targeted traits	Promoter for Cas9	Promoter for sgRNA	Method of transformation	References
Improvement in yield-related traits						
Rice	*Gn1a, DEP1, GS3 and IPA1*	Increase grain number, panicle structure, Size, plant architecture	ZmUbi	OsU6a	*Agrobacterium*	Li et al. (2016)
Wheat	*GASR7*	Increased thousand kernel weight	Ubi	TaU6	Particle bombardment	Zhang et al. (2016)
Rice	*GW2, GW5 and TGW6*	Grain weight	–	OsU3, OsU6, TaU3	*Agrobacterium*	Xu et al. (2016)
Rice	*OsCCD7*	Strigolactone biosynthesis	OsUbi	OsU3	*Agrobacterium*	Butt et al. (2018)
Improvement in quality traits						
Rice	*Waxy*	Reduced Amylose content	CaMV35S	OsU6-2	*Agrobacterium*	Zhang et al. (2018)
Rice	*OsNramp5*	Reduced Cadmium uptake	ZmUbi	OsU6, OsU3	*Agrobacterium*	Tang et al. (2017)
Rice	*rc*	Red pericarp	Ubi	U6	*Agrobacterium*	Zhu et al. (2019)
Rice	*SbeI, SbeIIb, SbeIIa*	Increased Amylose content	Ubi	OsU3	*Agrobacterium*	Sun et al. (2017)
Modification in biotic and abiotic stress responses						
Rice	*SAPK2*	Mediator of ABA signalling (Drought tolerance)	ZmUbi	OsU6	*Agrobacterium*	Lou et al. (2017)
Maize	*ARGOS8*	(Drought tolerance)	ZmUbi	ZmU6	Particle bombardment	Shi et al. (2017)
Wheat	*EDR1*	resistance to powdery mildew	ZmUbi	TaU6	Particle bombardment	Zhang et al. (2017)
Wheat	*TaMLO*	Mildew Resistance proteins	ZmUbi	TaU6	Particle bombardment	Wang et al. (2014)

(continued)

Table 3.1 (continued)

Crop	Target gene	Targeted traits	Promoter for Cas9	Promoter for sgRNA	Method of transformation	References
Rice	*eIF4G*	Translation initiation factor (Tungro virus resistance)	ZmUbi	TaU6	*Agrobacterium*	Macovei et al. (2018)
Rice	*OsERF922*	ERF transcription factor (Blast resistance)	ZmUbi	OsU6a	*Agrobacterium*	Wang et al. (2016)
Rice	*OsRR22*	Soil salinity	PUbi	OsU6a	*Agrobacterium*	Zhang et al. (2019)
Other traits:						
Wheat and barley	*Qsd1*	Increase seed dormancy	ZmUbi	OsU6	*Agrobacterium*	Abe et al. (2019)
Barley	*HvPM19*	Encode ABA-induced plasma membrane protein (Seed dormancy)	ZmUbi	TaU6	*Agrobacterium*	Lawrenson et al. (2015)
Rice, wheat	*OsACC-T1, TaDEP1, TaGW2*	Herbicide resistance[a]	Ubi-1	OsU3, TaU6	*Protoplast transfection*	Li et al. (2018)

Where Ubi—Ubiquitin; [a]Base editors were used instead of Cas9

are the expression level of sgRNA (single guide RNA) and *Cas9*, the secondary structure of sgRNA, GC content of the target DNA and codons of *cas9* (Ma et al. 2015). In order to overcome these problems and achieve high editing efficiency, the design of the construct needs further optimization. Generally *Agrobacterium* and particle bombardment have been used to introduce the sgRNA and Cas9 expression cassettes into target plants. The expression of sgRNA in plants is generally induced by RNA polymerase III promoters such as U3 and U6. Cas9 expression is generally driven by maize ubiquitin or cauliflower mosaic virus 35S promoters. Target genome sequences with high GC content increases the efficiency of editing as mutants can be detected more easily, but the secondary structure of sgRNA has a negative influence on genome editing (Liu et al. 2017).

The major problem faced in all genome editing methods is off-target effects. The Cas9 nuclease can tolerate some mismatches between the sgRNA and target sequences that help to prevent unwanted editing (Fu et al. 2013). Keeping this in mind, researchers are now designing specific sgRNA of 12 bp that can match only one site in genome (Doench et al. 2014). The requirement of specific PAM sequence to be present downstream of the target site desired to be edited also limits the scope of use of CRISPR/Cas9. Recently, Cas9 variants have been developed that show broad PAM compatibility while having higher target DNA specificity (Hu et al. 2018). Another modification of CRISPR/Cas system has been the use of base modifiers (cytosine and adenine deaminases) along with a deactivated Cas9 (dCAS9) or a CAS9 that causes a single-stranded nick (nCas9) instead of double-strand breakage. This approach has successfully been demonstrated in humans, rice, maize and wheat to bring about phenotypic changes through single base substitutions rather than insertions/deletions as in case of classical CRISPR/Cas9 (Li et al. 2018; Ren et al. 2018). In future, improvements are expected for enhancing specificity and fidelity and wider applicability of CRISPR-based technologies.

3.3 Application of CRISPR/Cas in Crop Improvement

Agriculture is facing many challenges due to various biotic and abiotic factors apart from greater emphasis on improving yield and quality aspects. CRISPR/Cas method of gene editing has been widely adopted in nearly 20 crop species so far (Ricroch et al. 2017). This includes rice, wheat, maize, soybean, barley, sorghum, potato, tomato, flax, rapeseed, *Camelina*, cotton, cucumber, lettuce, grapes, grapefruit, apple, oranges, and watermelon. The most frequent use of this editing tool is to generate gene knockouts/null alleles by insertion and deletions (Indels) leading to frameshift mutation or the introduction of a stop codon. Among the several crops edited using CRISPR/Cas, this chapter mainly focuses on cereal crops.

3.4 Modification of Yield Components

Cereals are the major food source worldwide. One trait of universal importance is yield. Yield is a quantitative trait governed by several genes and can be determined by several major traits like a number of tillers or panicles, number of grains per panicle and grain weight. In the case of rice, some of the genes that have been reported to be influencing yield-related traits are *IPA1* (Ideal Plant Architecture 1), *OsTB1* (Teosinte Branched 1), *Gn1a* (Grain Number 1a), *Ghd7* (Grains Height Date-7), *GS3* (Grain size), *GW2* (Grain weight), *GW5* (Grain weight), *OsSPL16* (Squamosa Promoter Binding Protein like 16) and *DEP1* (Dense and Erect Panicle). Genes *IPA1 and OsTB1* are responsible for tillering in rice (Miura et al. 2010). Number of grains per panicle is governed by genes like *Gn1a*, *Ghd7* (Li et al. 2013). Grain size is influenced by *GS3, GW2, GW5, OsSPL16* (Song et al. 2015; Wang et al. 2015), and *DEP1* controls the size of the panicle (Huang et al. 2009). Among the several genes reported for yield (Ikeda et al. 2013), the few that have been manipulated through CRISPR/Cas9 system are *Gn1a, DEP1, IPA1* and *GS3* (Li et al. 2016). The *Gn1a* allele harbours a mutation in *OsCKX2* gene, which encodes for cytokinin oxidase or dehydrogenase responsible for the degradation of active cytokinin. Mutation in this allele causes accumulation of cytokinin in the meristems enhancing the reproductive organs, ultimately leading to enhanced grain production (Ashikari et al. 2005). Similarly, a mutation in the miR156 cleavage site produced ideal plant architecture (IPA) with few unproductive tillers and more grains per panicle (Jiao et al. 2010). The mutant *DEP1* allele changed the inflorescence structure resulting in dense and erect panicles (Huang et al. 2009). *GS3* gene responsible for grain size contains OSR (Organ Size Regulation) domain in N terminal. Inactivation of this domain resulted in long grains and consequent increase in grain weight (Mao et al. 2010).

Grain weight is another important quantitative trait which supplements grain yield. Three major genes (*Grain Width2—GW2, Grain Width5—GW5, Thousand grain weight 6—TGW6*) negatively regulating weight were mutated simultaneously by CRISPR/Cas mediated multiplex editing resulting in the rapid generation and pyramiding of beneficial allele (Xu et al. 2016).

Strigolactones are novel carotenoid derived plant hormones that play a negative role in determining the number of tillers (Al-Babili and Bouwmeester, 2015) and induce germination of parasitic weeds. Therefore, manipulating genes for strigolactones biosynthesis will inhibit parasitic growth and increase yield (Conn et al. 2015). *OsCCD7* (*Carotenoid cleavage dioxygenase 7*) catalyses a key step in strigolactone biosynthesis, and a point mutation (C-T) in *OsCCD7* produces a dwarf and high tillering (*htd1*) phenotype (Zhang et al. 2011) Butt et al. (2018) used CRISPR/Cas9 for knocking out *Os*CCD7, which led to a drastic change in plant architecture and reduced strigolactones in root exudates.

3.5 Modifications for Quality Traits

Grain quality along with yield is the main target for all the plant breeders. In the past two decades, crop biofortification has become a major breeding objective to ensure nutritional security along with food security. The availability of genome sequences helps to facilitate gene discovery, targeted mutagenesis and also reveal functional aspects of grain quality traits. Conventional methods for grain quality assessment is lab oriented and time-consuming process (Cruz and Khush 2000), so a better understanding of the molecular basis of grain quality may lead to a faster and robust assessment method. Several regulatory genes, structural genes and chemical pathways are involved in determining grain quality and can be modified/edited by CRISPR/Cas system (reviewed by Fiaz et al. 2019).

CRISPR/Cas9 has been used to develop low gluten wheat lines by knocking out α-gliadin gene (Leon et al. 2018). The unique viscoelastic properties of wheat-derived foods are mainly due to the presence of the gluten proteins. The alpha-gliadin family is the main group associated with the development of coeliac disease as it affects more than 7% of the western populations (Sapone et al. 2011; Mustalahti et al. 2010). The alpha-gliadin in bread wheat is encoded by 100 genes and pseudogenes (Ozuna et al. 2015) and is organized in tandem at *Gli-2* loci of chromosomes

6A, 6B, 6D. The complexity of *Gli-2* locus and high copy number of alpha gliadin gene makes traditional and molecular breeding a tedious process for developing low gluten wheat. Through CRISPR/Cas, precise edition of the conserved targeted region in alpha-gliadin gene of both durum wheat and bread led to the production of low gliadin in seed kernels of wheat. Thus bread and durum wheat were developed with reduced immune reactivity to gluten-intolerant individuals.

Optimum amylose content in rice grain determines fluffiness and non-stickiness of cooked rice. Milled rice grains mostly contain starch which is composed of primarily to types of glucose polymers: amylase (unbranched) and amylopectin (branched). Amylose content varies from 0 to 30% in wild and cultivated varieties. Rice endosperm has at least three isoforms of *SBEs (Starch Branching Enzyme), SBEI, SBEIIa, SBEIIb* (Mizuno et al. 1992; Nakamura et al. 1992). *SBEIIb* is the most common isoform of *SBEII* present in rice grains whereas *SBEIIa* is expressed in leaves (Yamanouchi and Nakamura 1992; Ohdan et al. 2005; Yamakawa et al. 2007). Amylose content is increased by over-expression of suitable waxy allele (Itoh et al. 2003) or suppressing the expression of an enzyme involved in amylopectin synthesis (Morell and Myers 2005; Rahman et al. 2007). Initially, *sbeII* mutants were developed by physical and chemical treatments but they were accompanied by undesired and uncharacterized mutations. Finally, by designing a guide RNA specifically targeting the *SBEI* and *SBEIIb* genes in rice a homozygous transgene-free *SBEIIb* mutant with amylose content increase of 25% was obtained (Sun et al. 2017).

Conversely, very high amylose varieties become dry and hard after cooling and therefore, not preferred for consumption. Amylose content in rice is controlled by dominant Waxy gene which encodes a granule bound starch synthase (GBSS) leading to the production of amylose in the endosperm. Using CRISPR/Cas a loss of function of Waxy gene was generated resulting in reduced amylose content (Zhang et al. 2018).

Metal toxicity in grains is another important issue. Presence of excess cadmium in rice grain samples collected from Asian countries has been reported (Jallad 2015). Indica rice is reported to contain higher cadmium content in shoots and roots than the japonica types (Arao and Ae 2003; Uraguchi et al. 2009). Soil amelioration methods like soil removal and chemical washing are not able to limit the mobility and accumulation of Cd content to grains (Arao et al. 2010 and Huang et al. 2016). So, the development of low Cd uptake breeding lines is necessary to solve the problem. Physiologically, Cd is taken up by roots from the soil in the form of Cd^{2+} and transferred to shoots and when the plant reaches reproductive phase the stored Cd^{2+} from shoots is transferred to nodes and grains rather than mature leaves. Many Cd transporters like *OsIRT1*, *OsIRT2*, *OsNramp5* (NRAMP family proteins) and *OsNramp1* mediate Cd uptake through roots (Sasaki et al. 2012; Ishikawa et al. 2012). *OsHMA2* facilitates the xylem loading of Cd (Satoh–Nagasawa et al. Satoh-Nagasawa et al., 2012) and *OsLCT1* transfers cadmium from xylem to phloem (Uraguchi et al. 2011). Modifications in these metal transporter genes can reduce Cd accumulation in rice grains. Disruption of *OsNramp5* by CRISPR/Cas9 led to reduced Cd accumulation in *osnramp5* grains (Tang et al. 2017). This resulted in the production of grains with less than 0.05 mg/kg of Cd in contrast to 0.33 mg/kg to 2.90 mg/kg in wild type Indica line Huazhan.

Even though coloured rice like red, purple and brown pericarp varieties are available, most of the high yielding varieties available for consumption have white pericarp. Coloured rice contains more proanthocyanidins and anthocyanins which are health-promoting nutrients but the red colour is usually associated with wild-type undesirable traits like the presence of awn, shattering, low yield, etc. Rice red grain colour is produced by two genes *Rc* and *Rd* in wild rice *O. rufipogon*. Presence of 14 bp deletion in the seventh exon of *Rc* gene results in white grain. Using CRISPR/Cas technology functional reversion of recessive *rc* allele resulted in a successful conversion of white pericarp to red pericarp varieties without compromising the yield (Zhu et al. 2019).

In cereals, major work using CRISPR/Cas technology has been done in crops like rice, wheat and maize. Liu et al. (2019) reported the use of CRISPR/Cas in sorghum crop for modifying *CAD* (Cinnamyl alcohol dehydrogenase) and *PDS* (Phytoene desaturase) genes by particle bombardment. Generally, *CAD* gene encodes a key enzyme involved in lignin biosynthesis. Modification in lignin biosynthesis resulted in better sugar release and digestability (Fu et al. 2011).

3.6 Modifications Against Biotic Stress

Rice blast caused by *Magnaporthe oryzae* is a very common and serious problem faced by all the rice-growing countries. Even though several management practices are followed host plant resistance will be a better and more economical and effective method. Over the course of evolution, plants have developed resistance defense mechanisms against the pathogens. The plant hormones abscisic acid (ABA), jasmonic acid and ethylene play a major role in the defense (Rojo et al. 2003). ERF (Ethylene responsive factor) in plants are involved in multiple stress tolerance including biotic and abiotic stresses (Jung et al. 2007; Muller and Munne-Bosch 2015). During the period of infection, rice ERF genes *OsBIERF1*, *OsBIERF3* and *OsBIEF4* are expressed (Cao et al. 2006). Knocking down *OsERF922* gene by RNAi led to enhanced resistance to blast, suggesting that *OsERF922* is a negative regulator of blast disease resistance in rice (Liu et al. 2012). CRISPR/Cas approach targeting *OsERF922* led to various deletions and insertions (InDel) in the target site. The resulted mutant plants contained desired modifications excluding the transferred DNA (Wang et al. 2016). Similarly, knocking out of three homologs of *EDR1* (*Enhanced Disease Resistance*) gene in wheat by using CRISPR resulted in *Taedr1* plants with enhanced resistance to powdery mildew (Zhang et al. 2017). Targeted mutations were also successfully induced in three homoeoalleles that encode MLO (Mildew Resistance Locus) proteins. The induced mutation showed that three *TaMLO* homoeologs confer broad-spectrum resistance to powdery mildew (Wang et al. 2014).

Development of resistance against viral diseases is a major challenge as virus resistance sources are generally not available in crop germplasms. Rice tungro disease is a serious issue and is caused by the interaction between the Rice tungro spherical virus (RTSV) and Rice tungro bacilliform virus (RTBV). Resistance against RTSV was found in traditional cultivars and is reported to be governed by a recessive allele of *eIF4G* (*Translation Initiation factor 4 Gamma gene*). Single nucleotide polymorphisms were found by comparing the resistance and susceptible cultivars and the deletions in YVV amino acid residues were found to be associated with *eIF4G* allele conferring resistance. Mutations were created in susceptible variety IR 64 that possesses a susceptible S allele for *eIF4G,* whereas, the resistant varieties have non-S type allele with the deletion in YVV residues (Lee et al. 2010). CRISPR/Cas9 was used to create a novel allele of *eIF4G* carrying frameshift mutations especially adjacent to YVV residues and thereby conferring resistance (Macovei et al. 2018).

3.7 Modifications Against Abiotic Stress

CRISPR-based genome editing has been performed with great efficiency and precision for various abiotic stresses (reviewed by Jain 2015). Drought tolerance of a crop is measured by maintenance of yield under water insufficient conditions. In maize, higher yield under drought condition was obtained by reducing the sensitivity of maize plants to ethylene synthesis (Shi et al. 2015). *ARGOS* genes are negative regulators of ethylene response and are involved in signal transduction and enhanced drought tolerance (Guo et al. 2014). In general, the endogenous expression of *AGROS8* mRNA is low and non-uniform. Shi et al. (2015) reported that constitutive over-expression of *ARGOS8* produces more grains under drought stress condition. To achieve constitutive expression of *ARGOS8,* maize GOS2 promoter and 5s′ UTR with introns (GOS2Pro) were used to replace the native promoter present in *ARGOS8* gene. This was successfully achieved by using Cas9 endonuclease to create DSBs and integrating the GOS2 PRO into the upstream region of *ARGOS8* through HDR (Homology Directed Repair) mechanisms (Shi et al. 2017). This is a successful illustration of CRISPR/Cas9 being used for over-expression of a target gene.

Soil salinity is one of the important abiotic factors affecting crop production. Attempts are ongoing to understand the salt-tolerant mechanisms by altering biochemical pathways and enzymes. Rice grown in coastal wetland conditions is usually sensitive to salt stress. Several salt-tolerant genes are identified and cloned, such as *SKC1* (Shoot K^+ Concentration 1), *OsRR22* (Oryza sativa Response Regulator 22), *OsHAL3 (Oryza Sativa Halo tolerance protein), P5CS* (Pyrroline-5-Carboxylate Synthase), *SNAC2* (*Stress responsive NAC gene*), etc. The *SKC1* encodes a Na^+ transporter (Lin et al. 2004), *OsRR22* is responsible for cytokinin signal transduction and metabolism (Takagi et al. 2015), *OsHAL3* is a halotolerant protein, *P5CS* controls proline accumulation during salt stress (Strizhov et al. 1997) and *SNAC2* is a transcription factor gene controlling salt and cold tolerance (Hu et al. 2008). *OsRR22* encodes a 699 amino acid long B-type response regulator protein with N-terminal receiver domain and C terminal DNA binding domain (Takagi et al. 2015) and is in roots, stems and leaf sheaths involved in osmotic responses. Loss of function of *OsRR22* by using CRISPR/Cas9 technology resulted in increased salt tolerance (Zhang et al. 2019).

3.8 Modifications to Develop Male Sterility System

Male sterility is an important requirement for commercial hybrid seed production. Hybrid rice yields 10–20% more over conventional varieties. China is a leading country in hybrid seed production and uses mainly two-line (TGMS and PGMS) and three-line breeding methods. But a very limited number of germplasm lines have been identified that can be used as maintainer lines. For example, in China, only 1% of the rice germplasm can be used as maintainer lines (Deng et al. 2013). Two-line methods (PGMS and TGMS) express sterility in restricted environmental conditions. Nongken 58 s was the first PGMS line found in 1973 which is sterile under long-day conditions and found fertile under short-day conditions. Male sterility in PGMS lines is determined by *pms1, pms2* and *pms3* genes. *pms3* encodes a long noncoding RNA called long day-specific male fertility-associated RNA (LDMAR) (Ding et al. 2012). The first Indica TGMS line was identified in 1987 named as Annong S-1 (AnS-1) and the recessive gene *tms5* encodes an endonuclease RNase Z^{s1}. This RNase Z^{s1} controls the TGMS traits by degrading the temperature-sensitive ubiquitin fusion ribosomal protein L40 (Ub_{L40}) mRNA (Zhou et al. 2014, 2016). Initial attempts made using RNAi for targeting TMS5 gene knockouts led to incomplete depletion of the target gene. This problem was resolved by using CRISPR/Cas-based site-specific mutagenesis as a stable TGMS line with low CSIT (Critical Sterility Inducing Temperature) was generated ensuring the purity of hybrid seeds (Lei et al. 2014).

3.9 Accelerated Domestication

In several cases human selection and natural selection act in opposite directions. Domestication is a result of different selection processes that lead to increased adaptation or acclimatization of the crop plants and animals with respect to human requirements. Artificial selection or domestication by human activities has changed the evolution of the ecological niche. The domestication process has altered plant architectures at both genotypic and phenotypic level. These alterations resulted in the gradual transformation of the species from wild to elite lines and hybrids (Vaughan et al. 2007). With the help of available powerful tools like CRISPR/Cas, it is possible to alter or edit the genome and bring in desirable modifications. For example the gene responsible for precocious germination in wheat (*Qsd1*) was modified resulting in a longer dormancy period (Abe et al. 2019). Other than cereals the plant architecture in ornamentals, flower production in floricultural crops and fruit size in vegetables can also be altered. This technique holds the potential to accelerate domestication by improving these major traits (Lemmon et al. 2018).

3.10 Conclusion

Application of CRISPR/Cas technology widens the scope of crop improvement through genetic manipulation. The site-specific cleavage or site-directed mutagenesis provides more specificity and efficiency in the target site/gene manipulation. Presence of off-target effects is the main disadvantage in the gene-editing technology but attempts are ongoing to reduce these effects. CRISPR can act on exons or coding sequences and create null alleles, and also act on regulatory sequences and ORFs leading to enhanced expression. It can create single or multiple mutations either in homologous or nonhomologous regions. These advantages have led the scientists and plant breeders to focus on abiotic and biotic stress improvement, yield and quality of the grains etc., for introducing beneficial alleles across various crops. Another important advantage is non-transfer of transgenes to the next generation as they can be eliminated by the process of segregation leading to the production of homozygous transgene-free plants which can be selected and forwarded for future use. Therefore, this versatile technology promises to bring a new revolution in the field of crop improvement.

References

Abe F, Haque E, Hisano H et al (2019) Genome edited triple recessive mutation alters seed dormancy in wheat. Cell Rep 28:1362–1369

Al-Babili S, Bouwmeester HJ (2015) Strigolactones, a novel carotenoid-derived plant hormone. Annu Rev Plant Biol 66:161-186

Arao T, Ae N (2003) Genotypic variations in cadmium levels of rice grain. Soil Sci Plant Nutr 49:473–479

Arao T, Ishikawa S, Murakami M, Abe K (2010) Heavy metal contamination of agricultural soil and countermeasures in Japan. Paddy Water Environ 8(3):247–257

Ashikari M, Sakakibara H, Lin S et al (2005) Cytokinin oxidase regulates rice grain production. Science 309:741–745

Barrangou R, Fremaux C, Deveau H et al (2007) CRISPR provides acquired resistance against viruses in prokaryotes. Science 315:1709–1712

Butt H, Jamil M, Wang JY (2018) Engineering plant architecture via CRISPR/Cas9-mediated alteration of strigolactone biosynthesis. BMC Plant Biol 18:174

Campos H, Caligari PDS (2017) Genetic improvement of tropical crops. Springer International Publishing, Cham

Cao Y, Song F, Goodman RM, Zheng Z (2006) Molecular characterization of four rice genes encoding ethylene-responsive transcriptional factors and their expressions in response to biotic and abiotic stress. J Plant Physiol 163:1167–1178

Clarke JL, Zhang P (2013) Plant biotechnology for food security and bio economy. Plant Mol Biol 83:1–3

Conn CE, Bythell-Douglas R, Neumann D, et al (2015) Convergent evolution of strigolactone perception enabled host detection in parasitic plants. Science 349: 540–543

Cruz ND, Khush G (2000) Rice grain quality evaluation procedures. Aromatic rices. Oxford and IBH Publishing Co.Pvt.Ltd. New delhi, pp 15–28

Deng XW et al (2013) Hybrid rice breeding welcomes a new era of molecular crop design. Life Sci 864–868

Ding J, Lu Q, Ouyang Y et al (2012) A long noncoding RNA regulates photoperiod-sensitive male sterility, an essential component of hybrid rice. Proc Natl Acad Sci 109:2654–2659
Doench JG, Hartenian E, Graham DB et al (2014) Rational design of highly active sgRNAs for CRISPR-Cas9-mediated gene inactivation. Nat Biotechnol 32:1262
FAOSTAT (2016) FAOSTAT database. https://faostat3.fao.org/faostatgateway/go/to/download/Q/QC/E. Accessed 20 Sept 2019
Fiaz S, Ahmad S, Noor MA et al (2019) Applications of the CRISPR/Cas9 system for rice grain quality improvement: perspectives and opportunities. Int J Mol Sci 20:888
Fu C, Xiao X, Xi Y et al (2011) Downregulation of cinnamyl alcohol dehydrogenase (CAD) leads to improved saccharification efficiency in switchgrass. Bioenerg Res 4(3):153–164
Fu Y, Foden JA, Khayter C et al (2013) High-frequency off target mutagenesis induced by CRISPR-Cas nucleases in human cells. Nat Biotechnol 31:822–826
Georges F, Ray H (2017) Genome editing of crops: a renewed opportunity for food security. GM Crops Food 8:1–12
Guo M, Rupe MA, Wei J et al (2014) Maize ARGOS1 (ZAR1) transgenic alleles increase hybrid maize yield. J Exp Bot 65:249–260
Hsu PD, Scott DA, Weinstein JA et al (2013) DNA targeting specificity of RNA guided Cas9 nucleases. Nat Biotechnol 31:827–832
Hu H, You J, Fang Y et al (2008) Characterization of transcription factor gene SNAC2 conferring cold and salt tolerance in rice. Plant Mol Biol 67:169–181
Hu JH, Miller SM, Geurts MH et al (2018) Evolved Cas9 variants with broad PAM compatibility and high DNA specificity. Nature 556(7699):57–63
Huang S, Yang Y, Li Q et al (2016) Evaluation of the effects of lime-bassanite-charcoal amendment on the immobilization of cadmium in contaminated soil. Bull Environ Contam Toxicol 1–6
Huang X, Qian Q, Liu Z et al (2009) Natural variation at the DEP1 locus enhances grain yield in rice. Nat Gene 41:494–497
Ikeda M, Miura K, Aya K et al (2013) Genes offering the potential for designing yield-related traits in rice. Curr Opin Plant Biol 16:213–220
Ishikawa S, Ishimaru Y, Igura M et al (2012) Ion-beam irradiation, gene identification, and marker-assisted breeding in the development of low-cadmium rice. Proc Natl Acad Sci USA 109:19166–19171
Itoh K, Ozaki H, Okada K et al (2003) Introduction of Wx transgene into rice wx mutants leads to both high and low-amylose rice. Plant Cell Physiol 44:473–480
Jain M (2015) Function genomics of abiotic stress tolerance in plants: a CRISPR approach. Front Plant Sci 6:375
Jallad KN (2015) Heavy metal exposure from ingesting rice and its related potential hazardous health risks to humans. Environ Sci Pollut Res Int 22:15449–15458
Jiao Y, Wang Y, Xue D et al (2010) Regulation of OsSPL14 by OsmiR156 defines ideal plant architecture in rice. Nat Genet 42:541–544
Jung J, Won SY, Suh SC et al (2007) The barley ERF-type transcription factor HvRAF confers enhanced pathogen resistance and salt tolerance in Arabidopsis. Planta 225:575–588
Kim H, Kim JS (2014) A guide to genome engineering with programmable nucleases. Nat Rev Genet 15:321–334
Lawrenson T, Shorinola O, Stacey N et al (2015) Induction of targeted, heritable mutations in barley and Brassica oleracea using RNA-guided Cas9 nuclease. Genome Biol 16:258
Lee JH, Muhsin M, Atienza GA et al (2010) Single nucleotide polymorphisms in a gene for translation initiation factor (eIF4G) of rice (Oryza sativa) associated with resistance to rice tungro spherical virus. Mol Plant-Microbe Interact 23:29–38
Lei D, Tang W, Xie Z et al (2014) Solutions to insecurity problems in seed production of two-line hybrid rice. Agr Sci Technol 15:1160–1166
Lemmon ZH, Reem NT, Dalrymple J et al (2018) Rapid improvement of domestication traits in an orphan crop by genome editing. Nat Plants 4:766–770

Leon SS, Gil-Humanes J, Ozuna CV et al (2018) Low- gluten, nontransgenic wheat engineered with CRISPR/Cas9. Plant Biotechnol J 16:902–910

Li C, Zong Y, Wang Y et al (2018) Expanded base editing in rice and wheat using Cas9-adenosine deaminase fusion. Genome Biol 19:59

Li M, Li X, Zhou Z et al (2016) Reassessment of the four yield related genes Gn1a, DEP1, GS3 and IPA1 in rice using a CRISPR/Cas9 system. Front Plant Sci 7:377

Li S, Zhao B, Yuan D et al (2013) Rice zinc finger protein DST enhances grain production through controlling Gn1a/OsCKX2 expression. Proc Natl Acad Sci USA 110:3167–3172

Lin HX, Zhu MZ, Yano M et al (2004) QTLs for Na+ and K+ uptake of the shoots and roots controlling rice salt tolerance. Theor Appl Genet 108:253–260

Liu DF, Chen XJ, Liu JQ et al (2012) The rice ERF transcription factor OsERF922 negatively regulates resistance to Magnaporthe oryzae and salt tolerance. J Exp Bot 63:3899–3912

Liu G, Li J, Godwin ID (2019) Genome editing by CRISPR/Cas9 in sorghum through biolistic bombardent. Methods Mol Biol 1931:169–183

Liu H, Ding Y, Zhou Y et al (2017) CRISPR-P 2.0: an improved CRISPR-Cas9 tool for genome editing in plants. Mol Plant 10:530–532

Lou D, Wang H, Liang G et al (2017) OsSAPK2 confers abscisic acid sensitivity and tolerance to drought stress in rice. Front Plant Sci 8:993

Ma X, Zhang Q, Zhu Q (2015) A robust CRISPR/Cas9 system for convenient, high-efficiency multiplex genome editing in monocot and dicot plants. Mol Plant 8:1274–1284

Macovei A, Sevilla NR, Cantos C et al (2018) Novel alleles of rice eIF4G generated by CRISPR/Cas9-targeted mutagenesis confer resistance to rice tungro spherical virus. Plant Biotechnol J 16:1918–1927

Mali P, Yang L, Esvelt KM et al (2013) RNA-guided human genome engineering via Cas9. Science 339:823–826

Mao H, Sun S, Yao J et al (2010) Linking differential domain functions of the GS3 protein to natural variation of grain size in rice. Proc. Natl Acad Sci USA 107:19579–19584

Miura K, Ikeda M, Matsubara A et al (2010) OsSPL14 promotes panicle branching and higher grain productivity in rice. Nat Genet 42:545–549

Mizuno K, Kimura K, Arai Y et al (1992) Starch branching enzymes from immature rice seeds. J Biochem 112:643–651

Morell MK, Myers AM (2005) Towards the rational design of cereal starches. Curr Opin Plant Biol 8:204–210

Muller M, Munne-Bosch S (2015) Ethylene response factors: a key regulatory hub in hormone and stress signaling. Plant Physiol 169: 32–41

Mustalahti K, Catassi C, Reunanen A et al (2010) The prevalence of celiac disease in Europe: results of a centralized, international mass screening project. Ann Med 42:587–595

Nakamura Y, Takeichi T, Kawaguchi K, Yamanouchi H (1992) Purification of two forms of starch branching enzyme (Q-enzyme) from developing rice endosperm. Physiol Plant 84:329–335

Ohdan T, Francisco PB, Sawada T et al (2005) Expression profiling of genes involved in starch synthesis in sink and source organs of rice. J Exp Bot 56:3229–3244

Ozuna CV, Iehisa JCM, Gimenez MJ et al (2015) Diversification of the celiac disease a gliadin complex in wheat: a33 merpeptide with six overlapping epitopes, evolved following polyploidization. Plant J 82:794–805

Qi LS, Larson MH, Gilbert LA (2013) Repurposing CRISPR as an RNA-Guided platform for sequence specific control of gene expression. Cell 152:1173–1183

Rahman S, Bird A, Regina A et al (2007) Resistant starch in cereals: exploiting genetic engineering and genetic variation. J Cereal Sci 46:251–260

Ren B, Yan F, Kuang Y et al (2018) Improved Base editor for efficiently inducing genetic variations in rice with CRISPR/Cas9-guided hyperactive hAID mutant. Mol Plant 11(4):623–626

Ricroch A, Clairand P, Harwood W (2017) Use of CRISPR systems in plant genome editing toward new opportunities in agriculture. Emerg Top Life Sci 1:169–182

Rojo E, Solano R, Sánchez-Serrano JJ (2003) Interactions between signaling compounds involved in plant defense. J Plant Growth Regul 22:82–98

Sapone A, Lammers KM, Casolaro V et al (2011) Divergence of gut permeability and mucosal immune gene expression in two gluten-associated conditions: celiac disease and gluten sensitivity. BMC Med 9:23

Sasaki A, Yamaji N, Yokosho K, Ma JF (2012) Nramp5 is a major transporter responsible for manganese and cadmium uptake in rice. Plant Cell 24:2155–2167

Satoh-Nagasawa N, Mori M, Nakazawa N, et al (2012) Mutations in rice (Oryza sativa) heavy metal ATPase 2 (OsHMA2) restrict the translocation of zinc and cadmium. Plant Cell Physiol 53: 213–224

Shi J, Gao H, Wang H et al (2017) ARGOS8 variants generated by CRISPR-Cas9 improve maize grain yield under field drought stress conditions. Plant Biotechnol J 15(2):207–216

Shi J, Habben JE, Archibald RL et al (2015) Overexpression of ARGOS genes modifies plant sensitivity to ethylene, leading to improved drought tolerance in both Arabidopsis and maize. Plant Physiol 169:266–282

Song XJ, Kuroha T, Ayano M et al (2015) Rare allele of a previously unidentified histone H4 acetyltransferase enhances grain weight, yield, and plant biomass in rice. Proc Natl Acad Sci USA 112:76–81

Strizhov N, Abraham E, Okresz L et al (1997) Differential expression of two P5CS genes controlling proline accumulation during salt stress requires ABA and is regulated by ABA1, ABI1 and AXR2 in Arabidopsis. Plant J 12:557–569

Sun Y, Jiao G, Liu Z et al (2017) Generation of high amylose rice through CRISPR/Cas9 mediated targeted mutagenesis of starch branching enzymes. Front Plant Sci 8:298

Takagi H, Tamiru M, Abe A et al (2015) Mutmap accelerates breeding of a salt tolerant rice cultivar. Nat Biotechnol 33:445–449

Tang L, Mao B, Li Y et al (2017) Knockout of OsNramp5 using the CRISPR/Cas9 system produces low Cd-accumulating indica rice without compromising yield. Sci Rep 7:14438

Uraguchi S, Kamiya T, Sakamoto T et al (2011) Low-affinity cation transporter (OsLCT1) regulates cadmium transport into rice grains. Proc Natl Acad Sci 108(52):20959–20964

Uraguchi S, Mori S, Kuramata M et al (2009) Root-to-shoot Cd translocation via the xylem is the major process determining shoot and grain cadmium accumulation in rice. J Exp Bot 60:2677

Vaughan DA, Balzs E, Harrison JSH (2007) From crop domestication to super-domestication. Ann Bot 100:893–901

Wang F, Wang C, Liu P et al (2016) Enhanced rice blast resistance by CRISPR/Cas9—targeted mutagenesis of the ERF transcription factor gene OsERF922. PLoS ONE 11:e0154027

Wang S, Li S, Liu Q et al (2015) The OsSPL16-GW7 regulatory module determines grain shape and simultaneously improves rice yield and grain quality. Nat Genet 47:949–954

Wang Y, Cheng X, Shan Q et al (2014) Simultaneous editing of three homoeoalleles in hexaploidy bread wheat confers heritable resistance to powdery mildew. Nat Biotechnol 32:947–951

Xu R, Yang Y, Qin R et al (2016) Rapid improvement of grain weight via highly efficient CRISPR/Cas9 mediated multiple genome editing in rice. J Genet Genomics 43(8):529–532

Yamakawa H, Hirose T, Kuroda M, Yamaguchi T (2007) Comprehensive expression profiling of rice grain filling-related genes under high temperature using DNA microarray. Plant Physiol 144:258–277

Yamanouchi H, Nakamura Y (1992) Organ specificity of isoforms of starch branching enzyme (Q-enzyme) in rice. Plant Cell Physiol 33:985–991

Zhang A, Liu Y, Wang F et al (2019) Enhanced rice salinity tolerance via CRISPR/Cas9 targeted mutagenesis of the OsRR22 gene. Mol Breed 39:47

Zhang B, Tian F, Tan L, Xie D, Sun C (2011) Characterization of a novel high-tillering dwarf 3 mutant in rice. J Genet Genomics 38(9):411–418

Zhang J, Zhang H, Botella JR, Zhu JK (2018) Generation of new glutinous rice by CRISPR/Cas9—targeting mutagenesis of the Waxy gene in elite rice varieties. J Integr Plant Biol 60:369–375

Zhang Y, Bai Y, Wu G et al (2017) Simultaneous modification of three homoeologs of TaEDR1 by genome editing enhances powdery mildew resistance in wheat. Plant J 91:714–724

Zhang Y, Liang Z, Zong Y et al (2016) Efficient and transgenefree genome editing in wheat through transient expression of CRISPR/Cas9 DNA or RNA. Nat Commun 7:12617

Zhou H, He M, Li J et al (2016) Development of commercial thermo sensitive genic male sterile rice accelerates hybrid rice breeding using the CRISPR/Cas9 mediated TMS5 editing system. Sci Rep 6:37395

Zhou H, Zhou M, Yang Y et al (2014) RNase Z (S1) processes UbL40 mRNAs and controls thermosensitive genic male sterility in rice. Nat Commun 5:4884

Zhu Y, Lin Y, Chen S et al (2019) CRISPR/Cas9 mediated functional recovery of the recessive rc allele to develop red rice. Plant Biotechnol J. https://doi.org/10.1111/pbi.13125

Chapter 4
Improvement of Floriculture Crops Using Genetic Modification and Genome Editing Techniques

Ayan Sadhukhan and Heqiang Huo

Abstract Floriculture is an integral part of modern agriculture. Genetic transformation techniques pave way for the development of biotechnology assisted novel varieties which are appealing to customers. Customer preference in floriculture changes rapidly with time. Recently discovered CRISPR/Cas genome editing technique is useful in editing key genes attributing to a trait with the enhanced aesthetic appeal of plants. This chapter primarily discusses the application of biotechnology-based tools for improvement of floriculture plants to keep up with consumer appeal.

Keywords Floriculture · CRISPR/Cas · Foliage plants · *Antirrhinum* · *Chrysanthemum* · *FOREVER YOUNG FLOWER* gene

4.1 Introduction

Floriculture is an important sector of the agriculture industry, which is comprised of ornamental plants for cut flowers, home gardening, indoor and outdoor landscaping. Characteristics such as flower colour and foliage shape contribute to the aesthetic values of ornamental plants. With the advancement in genetic transformation methods, there has been a boom in the efforts to improve desirable traits of ornamental plants in a comprehensive scientific manner. Both structural and regulatory genes can be introduced in ornamental plants for improving their aesthetic values through modifying foliage or floral characteristics, elongating shelf life as well as enhancing environmental stress tolerance. Although only a few genetically engineered ornamental plants have been released in the market (Tanaka and Brugliera 2013; www.florigene.com), yet the efforts for genetic manipulation of ornamentals via genetic modification (GM) and genome editing (GE) continues in laboratories worldwide. In this chapter, we discuss the recent progress, mostly in the past five years, in the genetic improvement of ornamental plants and the challenges that lie ahead, some of which, can be addressed via GE.

A. Sadhukhan · H. Huo (✉)
Mid Florida Research and Education Centre, University of Florida, Apopka, FL, USA
e-mail: hhuo@ufl.edu

A. Bhattacharya et al. (eds.), *CRISPR/Cas Genome Editing*, Concepts and Strategies in Plant Sciences, https://doi.org/10.1007/978-3-030-42022-2_4

4.2 Agrobacterium-Mediated Transformation of Ornamental Plants

Agrobacterium tumefaciens-mediated natural gene transfer to plants through its tumour-inducing (Ti) plasmid was one of the ingenious discoveries of the twentieth century and now a widely used tool for genetic engineering of many models and crop plants. In routine applications, disarmed Ti plasmid vectors, with their tumour-inducing genes removed, carrying a gene of interest downstream of a constitutive or tissue-specific or inducible promoter in their T-DNA, is mobilised to a hypervirulent strain of *Agrobacterium*. After bacterial infection to a wounded part of the plant, the bacterium delivers its T-DNA to the plant cell for subsequent random integration into the plant genome. Transgenic plants are generated either through direct organogenesis or somatic embryogenesis from differentiated tissue, that is the infected plant material/explant, or indirectly through a de-differentiated disorganised mass of cells (callus). Additional selectable marker genes like antibiotic/herbicide resistance genes in the T-DNA allow for the selection of transformed plants. Ornamental plant genotype and choice of explants determine the feasibility of stable gene transfer and subsequent regeneration of transgenic plants. Woody ornamentals present barriers to regeneration due to phenolic compounds and most monocot ornamentals cannot get infected by *Agrobacterium* requiring other approaches of DNA delivery. In recent years, optimised protocols for *Agrobacterium*-mediated genetic transformation of several important ornamental plants have been reported (Table 4.1). For example, an optimal combination of plant growth regulators, 6-benzyladenine and 3-indole-acetic acid in Murashige and Skoog (MS) medium (Murashige and Skoog 1962), led to efficient direct shoot regeneration under kanamycin antibiotic selection from petiole explants after *Agrobacterium* infection of gerbera (*Gerbera hybrid* Hort.) cultivar 'Gold Eye' (Olsen et al. 2015). Although several explants, viz. shoot, leaves, bulbs, floral stalks etc. have been used for genetic transformation of orchids (Teixeira da Silva et al. 2016), recently a protocol was developed for direct organogenesis from flower petals of the orchid *Dendrobium* Sonia 'Earsakul'. ½-MS media supplemented with 1-naphthaleneacetic acid (NAA) and benzylaminopurine was used to regenerate meristemoids from *Agrobacterium*-infiltrated petals (Sahagun et al. 2018). A specialised medium for orchids, New Dogashima medium (Belarmino and Mii 2000), with maltose, NAA and benzyladenine was used for *Agrobacterium*-mediated transformation of protocorms of *Phalaenopsis* orchid. The transformed reporter gene *eGFP* was stably inherited in the progeny when the transgenic plants were back-crossed to several *Phalaenopsis* orchid varieties (Hsing et al. 2016). Recalcitrance to genetic transformation is observed in many cultivated crops including some ornamental plants. The highly stress-tolerant and large-flowered Korean chrysanthemum cultivar Shinma is recalcitrant to *Agrobacterium*-mediated transformation. In a transformation protocol optimisation study, co-cultivation temperature and *Agrobacterium* strain were found to be the key determinants of transformation success in this cultivar (Naing et al. 2016). In *Gladiolus hybridus* cv. 'Advance Red' cormel slice explants were precultured and co-cultivated with *A. tumifaciens* strain GV3101

Table 4.1 Genetic engineering of ornamental plants in recent years

Plant species	Cultivar	Gene transformed	Nature of transformation	Explant	Method of gene transfer	Phenotype	References
Dendrobium Sonia	Earsakul	*gus*	Overexpression	Petal	*Agrobacterium*	GUS expression	Sahagun et al. (2018)
Petunia hybrida	Creepia White, Dressup Neo	*eYGFPuv*	Overexpression	Leaf disk	*Agrobacterium*	Fluorescent flowers and leaves visible to naked eye	Chin et al. (2018)
Pelargonium zonale	Katinka	*A. thaliana etr1-1* mutant allele	Floral and senescence-specific expression	Petiole from shoot apex	*Agrobacterium*	Reduced senescence upon external ethylene exposure	Gehl et al. (2018)
Phalaenopsis aphrodite	TH274-1, Ama (diploid and tetraploid)	*eGFP, hptII*	Overexpression	Protocorm	*Agrobacterium*	GFP expression	Hsing et al. (2016)
Chrysanthemum	Shinma	*RsMYB1, bar*	Overexpression, petal-specific expression	Leaf	*Agrobacterium*	Transgene expression, Basta resistance	Naing et al. (2016)
Torenia fournieri	Crown violet	*AtTCP3*-SRDX	Flower specific repession of gene expression	Leaf disk	*Agrobacterium*	Changes in flower colour and shape	Sasaki et al. (2016)
Petunia x*hybrida*	Mitchell diploid	*PhMLO1*	Silencing (RNAi)	Leaf	*Agrobacterium*	Resistance to Powdery mildew (*Podoshaera xanthii*)	Jiang et al. (2016)

(continued)

Table 4.1 (continued)

Plant species	Cultivar	Gene transformed	Nature of transformation	Explant	Method of gene transfer	Phenotype	References
Lilium longiflorum	Nellie White	*Oc-IDD86*	Overexpression	Suspension cells, callus and bulb scales	Gene gun	Resistance to the root nematode *Pratylenchus penetrans*	Vieira et al. (2015)
Rosa hybrida	Glad Tidings	*gus*	Overexpression	Embryogenic callus	Gene gun	GUS expression	Dhanya and Thangavel (2015)
Rosa hybrid Nierembergia sp.	WKS124 NB18	Pansy *F3′5′H* and *Torenia A3′5′OMT*	Overexpression	Stem intrnode, shoot apex	*Agrobacterium*	Higher accumulation of methylated anthocyanins an magenta colour	Nakamura et al. (2015)
Petunia × hybrida	V26	AroG*	Overexpression	Leaf	*Agrobacterium*	Higher phenylalanine and frangrance levels in flowers	Oliva et al. (2015)
Dendranthema morifolium	–	*DmCPD, DmGA20ox*	Silencing (RNAi)	–	*Agrobacterium*	Reduced brassinosteroid and gibberelin, miniature plants	Xie et al. (2015)
Chrysanthemum morifolium	Zhongshanzigui, Jinba	*CmWRKY17*	Overexpression	Leaf disk	*Agrobacterium*	Salt sensitivity	Li et al. (2015)

(continued)

Table 4.1 (continued)

Plant species	Cultivar	Gene transformed	Nature of transformation	Explant	Method of gene transfer	Phenotype	References
Rosa multiflora	–	*RhMLO1*	Silencing	–	*Agrobacterium*	Resistance to Powdery mildew (*Podoshaera pannosa*)	Qiu et al. ((2015)
Gladiolus sp.	Peter pears	D4E1(synthetic)	Overexpression	–	Particle bombardment	Resistance to *Fusarium oxysporum*	Kamo et al. (2015)
Lilium sp.	Star gazer	*RCH10*	Overexpression	Bulblet basal plate disk	*Agrobacterium*	Resistance to *Botrytis cinerea*	Núñez de Cáceres González et al. (2015)
Chrysanthemum morifolium	Shuho-no-chikara	*cry1Ab, sarcotoxin IA*	Overexpression	Shoot tip	*Agrobacterium*	Resistance against whte rust (*Puccinia horiana*), lepidopteran insect (*Helicoverpa armigera*)	Shinoyama et al. (2015)
Lilium longiflorum	Nellie White	*uidA, nptII*	Overexpression	Suspension cells, callus and bulb scales	Gene gun	GUS expression	Kamo (2014)
Gladiolus hybridus	Advance Red	*gus, hpt*	Overexpression	Cormel	*Agrobacterium*	GUS expression	Wu et al. (2014)

(continued)

Table 4.1 (continued)

Plant species	Cultivar	Gene transformed	Nature of transformation	Explant	Method of gene transfer	Phenotype	References
Lilium tenuifolium Oriental × Trumpet	Robina	*gus*	Overexpression	Embryogenic cell suspension culture	*Agrobacterium*	GUS expression	Qi et al. (2014)
Gladiolus hybridus	Roses supreme	*GhPDS*	Silencing	Cormel	VIGS	Leaf photobleaching	Zhong et al. (2014)
Petunia axillaris × (*P. axillaris* × *P. hybrida*) *Eustoma grandiflorum*	Mitchell	*ROSEA1*	Overexpression	Leaf disk	*Agrobacterium*	Enhanced anthocyanin pigmentation	Schwinn et al. (2014)
Iris germanica	Fire bride	*crtB* from *Pantoea agglomerans*	Flower specific expression	Suspension culture	*Agrobacterium*	Orange and pink colouration in ovaries, anthers and flower stalks	Jeknić et al. (2014)
Torenia fournieri	Crown white	*CpYGFP*	Overexpression	Leaf disk	*Agrobacterium*	Fluorescent flowers	Sasaki et al. (2014)
Chrysanthemum crassum	Jinba	*CcSOS1*	Overexpression	Young leaf	*Agrobacterium*	Salinity tolerance	An et al. (2014)
Chrysanthemum crassum	Jinba	*CcSOS1, CdICE1*	Overexpression	Young leaf	*Agrobacterium*	Salt, cold, drought tolerance	Song et al. (2014)
Chrysanthemum morifolium	Nannongyinshan	*CmPT1*	Overexpression	Leaf segments	*Agrobacterium*	Low phosphate tolerance	Liu et al. (2014)
Petunia × *hybrida*	Primetime Blue, Mitchell diploid	*PhHD-ZIP*	Silencing, overexpression	Seedlings	VIGS, *Agrobacterium*	Delayed flower senescence, accelerated flower senescence	Chang et al. (2014)

harbouring a pCAMBIA series binary vector carrying a reporter *ß-glucuronidase* (*gus*) gene under the constitutive cauliflower mosaic virus (CaMV) *35S* promoter. Treatment with *N*-phenyl-*N*′-1,2,3-thiadiazol-5-yl urea led to callus induction from infected explants and subsequent somatic embryogenesis (Wu et al. 2014). Somatic embryogenesis protocols from *Agrobacterium*-transformed explants of important ornamental plants, carnation (Iantcheva 2015) and rose (Shen et al. 2016), have been reviewed earlier. A protocol was developed for producing embryogenic cell suspension cultures from filaments and stem axis of ornamental lily (*Lilium tenuifolium* oriental × trumpet 'Robina') amenable for *Agrobacterium*-mediated genetic transformation and subsequent plant regeneration. MS media with optimised concentrations of picloram, NAA, and thidiazuron was used for callus induction and suspension cultures, and regeneration was carried out in hormone-free MS (Qi et al. 2014).

4.3 Biolistic Gene Delivery to Ornamental Plants

A gene gun is an apparatus that bombards gold-coated plasmid DNA directly to plant cells assisted by helium gas for genetic transformation, particularly useful in plants resistant to *Agrobacterium* infection, for example monocots. PDS-1000/Helium system (Bio-Rad, Richmond, CA) gene gun was used to bombard plasmid DNA into cultured cells, calli and bulb scales of monocot ornamental *Lilium longiflorum* 'Nellie White' (Kamo 2014; Vieira et al. 2015). Embryogenic calli of dicot ornamental species like rose have been also been used for gene gun mediated DNA bombardment for successful genetic transformation (Dhanya and Thangavel 2015). Optimisation of bombardment rupture-disk pressure and post-bombardment selection media greatly determined regeneration and transformation efficiency in these studies.

4.4 Virus-Induced Gene Silencing

Virus-induced gene silencing (VIGS) is a tool of reverse genetics in which modified viral vectors are used to carry a portion of a target gene to produce a double-stranded RNA that directs silencing of the target gene for studies of gene function. The viral vector with a gene of interest is inoculated into the plant using several techniques including *Agrobacterium*-infiltration assisted by vacuum. Other viral proteins are supplied by another assisting vector for viral assembly inside the plant. This leads to the fast appearance of the phenotype without the need for plant genomic transformation (Unver and Budak 2009). To efficiently silence the *phytoene desaturase* ortholog *GhPDS* in *Gladiolus hybridus*, a tobacco rattle virus-based VIGS vector carrying a conserved region of the gene was transferred to *Gladiolus* cormels and young plants via *Agrobacterium*-infiltration, leading to leaf photobleaching phenotype (Zhong et al. 2014).

4.5 Gene Identification from Ornamental Plants

Functionally characterised genes from model plants are routinely used in genetic transformation of ornamental plants for manipulation of various traits. Orthologs of these genes from ornamental plants have also been isolated by a degenerate-polymerase chain reaction and rapid amplification of cDNA ends, and functionally validated in model systems or in *cis*-genic ornamentals. In recent years functional genomics, particularly by way of transcriptomic analysis through next-generation sequencing, have been initiated in ornamental plants for identification of novel candidate genes critically regulating desirable traits. We shall discuss some recent examples of gene identification from ornamental species.

A novel cytochrome P450 gene *CYP76AD6* was isolated from beet (*Beta vulgaris*) and four o'clock (*Mirabilis jalapa*) and expressed in tobacco to form red betalain pigment. The enzyme encoded by this gene caused hydroxylation of tyrosine to L-3,4-dihydroxyphenylalanine, an early step in betalain biosynthesis and resulted in red-pigmented tobacco (Polturak et al. 2016). Similarly, five betalain biosynthesis genes were isolated from the plant *Amaranthus tricolor* which produces deep purple flowers and the gene expression patterns in different tissues in response to different phytohormones (Zheng et al. 2016). This opens ways for future engineering of pigments to increase the value of different ornamentals. Functional studies of photosynthesis in the highly valuable ornamental orchid, *Phalaenopsis aphrodite*, was conducted by biochemical and molecular approaches (Ping et al. 2018). By precise measurements of diurnal CO_2 exchange and malate levels, phosphoenolpyruvate carboxylase activity and gene expression levels during tissue culture it was determined that photosynthesis shifted from C3 to crassulacean acid metabolism (CAM) during early development of this orchid. This opens up ways for efficient in vitro propagation and genetic engineering for improving the photosynthetic performance of orchids. Shoot branching is an important factor in determining plant shape for both flowering and non-flowering ornamentals. Function of a chrysanthemum ortholog of shoot-branching regulating *D27* gene from Arabidopsis, reportedly involved in strigolactone biosynthesis, was validated by complementation of Arabidopsis *d27-1* mutant (Wen et al. 2016). *DgD27* was auxin-induced in the shoots and repressed upon decapitation and dark treatment. This gene was also found to be highly responsive to phosphate (P) deficiency and modulate levels of different phytohormones in the shoot. Thus, *DgD27* was identified as a candidate for future applications in engineering shoot architecture in ornamentals. A sterol glycosyl transferase ortholog (*WuSGTL1*) isolated from Aswagandha or Indian ginseng (*Withania somnifera*) led to higher accumulation of glycosylated sterols and reactive oxygen species (ROS) detoxifying metabolites and enzymes in transgenic tobacco which were more resistant to both biotic and abiotic stress (Pandey et al. 2014). Lee et al. (2018) characterised the expression of *Phaius tankervilliae 9-cis-epoxycarotenoid dioxygenase 1 (PtNCED1)*, regulating the biosynthesis of abscisic acid and the rate of seed germination of this ornamental orchid for future use in genetic engineering.

The transcriptome of the scented tropical ornamental plant *Hedychium coronarium* was analysed by RNA sequencing using the Illumina platform. This study could narrow down flower-specific genes controlling petal development and involved in the biosynthesis of floral terpenes and benzoids. The function of some genes was characterised further by analysis of the aromatic volatiles (Yue et al. 2015). To identify the molecular basis of extreme drought adaptation of the orchid *Dendrobium wangliangii*, transcriptome analysis was conducted resulting in the unravelling of metabolic processes and candidate genes responsible for drought tolerance (Zhao et al. 2019). Potential candidate genes and novel isoforms of genes regulating vernalisation (cold treatment necessary for flowering) were identified in lily (*Lilium longiflorum* cv. White Heaven) by RNA sequencing (Villacorta-Martin et al. 2015). Transcriptome of leaves and flower buds of *Gerbera hybrida* was analysed by RNAseq and transcripts regulating the biosynthesis of floral pigments and phytohormones, and signal transduction for disease resistance were studied in detail in silico (Fu et al. 2016).

4.6 Genetic Engineering for Improvement of Ornamental Traits

4.6.1 Engineering of Floral Traits

4.6.1.1 Flower Colour

Flowering ornamentals are the most valued. Considerable efforts have always been guided to modify important floral traits like flower shape, colour and fragrance, as well as flowering time in many ornamental plant species. Flavonoids and their coloured derivatives, anthocyanins are vacuolar pigments which impart various bright flower colours due to their different pH, viz. magenta, blue and purple. Carotenoids are other pigments which impart orange, red and pink colours (Jeknić et al. 2014). Genetic engineering by introducing transgenes to enhance or modify anthocyanin accumulation remains one of the most prevalent strategies to alter flower colour. Overexpression of a R2R3-MYB transcription factor (TF) gene *ROSEA1* from *Antirrhinum majus* in petunia increased anthocyanin accumulation. Expression levels of genes of flavonol biosynthesis pathway, viz. chalcone synthase and anthocyanidin synthase were upregulated in the flower petals of the overexpressor lines. These lines were also crossed with another transgenic overexpressing maize basic helix-loop-helix TF gene *LEAF COLOR*, which enhances foliage pigmentation when exposed to high light by regulating flavonoid related genes (Schwinn et al. 2014). The enhanced visible pigmentation phenotypes were also tested under field conditions. Thus, evidently, pigmentation in both flower and vegetative tissues can be engineered together in ornamentals by genetic engineering. In another instance, *S-adenosylmethionine: anthocyanin 3′,5′-O-methyltransferase* gene was isolated from

Torenia hybrida petals that accumulate malvidin type anthocyanins and introduced into rose which doesn't contain methylated anthocyanins. GM rose and cupflower (*Nierembergia* sp.) petals expressed brilliant magenta petal colour due to increased accumulation of malvidin and other methylated anthocyanins (Nakamura et al. 2015). Metabolic engineering is another approach to engineer changes in flower colour through changes in the synthesis of pigments. For example, ectopic expression in *Iris germanica* of a phytoene synthase gene from the bacterium *Pantoea agglomerans* under the control of the promoter region of a gene for capsanthin–capsorubin synthase from *Lilium lancifolium* was attempted to change flower colour by shifting the metabolism from anthocyanin to carotenoid synthesis. This was met with partial success as certain flower parts showed colour changes from green/white to pink (Jeknić et al. 2014).

Cytochromes P450 catalyse various reactions in flavonoid/anthocyanin biosynthesis, e.g. flavonoid 3′-hydroxylase, flavonoid 3′,5′-hydroxylase and flavone synthase II, determining flower colour. The degree of hydroxylation on the B-ring of anthocyanins determine the blueness of flower colour. Flavones are co-pigments which bring about additional blueness. GM roses, chrysanthemums and carnations producing novel blue and violet coloured flowers have been successfully engineered by introducing the genes coding first two enzymes in which they are deficient. On the other hand, downregulation of these cytochromes P450 genes led to redder flowers in GM torenia and petunia (Tanaka and Brugliera 2013; Sasaki and Nakayama 2015). To aid this type of floral genetic engineering avoiding undesirable effects of the CaMV *35S* promoter on other plant organs, researchers have tried to develop flower-specific promoters. Chimeric promoters produced by fusing *octopine synthase* enhancer with *chalcone synthase A* core promoter from petunia and an additional Ω enhancer showed high corolla specificity in the expression of reporter *gus* gene in GM torenia (Du et al. 2014). Chimeric repressors are short 12-amino acid repressor domains, for example SRDX from Arabidopsis, fused to a transcriptional activator to convert it into a strong transcriptional repressor (Hiratsu et al. 2003). SRDX fused to Arabidopsis *TEOSINTE BRANCHED1, CYCLOIDEA, and PCF transcription factor 3* (*TCP3*) were expressed under the control of flower-specific promoters from torenia and found to produce novel paler purple flower colours and shapes in GM torenia. The TCP3-SRDX repressor caused suppression of anthocyanin biosynthesis genes causing different paler shades of flower colour and changes in epidermal cell shapes reflected in altered petal shapes. The flower-specific promoters could overcome the defects in leaves and stems found when using the CaMV *35S* promoter (Sasaki et al. 2016).

4.6.1.2 Fragrance

Fragrance is one of the most sought-after ornamental traits and metabolic engineering has been successfully applied to enhance the fragrance of some flowers. For example, petunia plants were transformed with a bacterial gene encoding 3-deoxy-diarabino-heptulosonate 7-phosphate synthase enzyme of the shikimate pathway to increase flower fragrance by the formation of aromatic amino acids. The GM plants produced high phenylalanine levels in flower petals. Phenylalanine is a precursor of fragrant volatile benzenoid-phenylpropanoids which also emitted increasingly from the GM flowers. The transgene expression had no effect on flower colour producing pigments or flower lifetime. Even phenylalanine produced by the transgene expressed in the leaves in some lines were transported to the floral parts to produce increased fragrance (Oliva et al. 2015).

4.6.1.3 Male Sterility

Male sterility is a desired trait in some cultivated plants including ornamentals for ease in production of hybrids or to prevent horizontal gene transfer from transgenic plants via pollen to wild/related species. Genetic engineering is sometimes applied to induce male sterility. Overexpression of ethylene receptor genes *CmETR1/H69A* from melon caused temperature-sensitive male sterility in chrysanthemum (*C. morifolium* Ramat.) (Shinoyama et al. 2012). The transgenic lines did not produce pollen at 20–35 °C but within 10–15 °C normal pollen was produced due to suppression of the transgene at the low temperatures. More research is necessary to optimise the unwanted temperature effects on engineering male sterility.

4.6.2 Engineering of Fluorescent Flowers and Plants

Genetic engineering has led to the generation of fluorescent plants both as tools for fundamental research as well as for ornamental purposes. A *yellowish-green fluorescent protein* (*CpYGFP*) gene from the marine plankton *Chiridius poppei* fused to CaMV *35S* promoter and Arabidopsis *alcohol dehydrogenase* 5′-untranslated region and a *heat shock protein 18.2* terminator sequence were introduced in the ornamental plant torenia. The transgenic plants expressed bright fluorescence which was visible macroscopically using simple excitation lights and emission filters (Sasaki et al. 2014). With some more improvisation by using *eYGFPuv*, a derivative of *CpYGFP* which has a stronger excitation under UV (398 nm) and a longer Stokes shift (104 nm), and a different 5′-UTR from Arabidopsis *cold regulated 47* gene, GM petunia plants were generated which produced stable green fluorescence visible to the naked eye (Chin et al. 2018).

4.6.3 Engineering Shoot Architecture

Shoot size and branching are regulated by different phytohormones and genetic manipulation of such phytohormones can alter plant architecture, particularly important in case of ornamental plants. Shoot architecture-regulating genes *DmCPD* and *DmGA20ox*, were cloned from chrysanthemum (*Dendranthema morifolium*) and both genes were silenced by RNAi to produce miniature GM chrysanthemum. The miniaturisation could be attributed to the suppression of genes related to brassinosteroid and gibberellin metabolism and the concomitant decrease in these phytohormones in the transgenic plants (Xie et al. 2015). The flowering ornamental plant *Mecardonia* sp. was transformed with wild *Agrobacterium rhizogenes* in a novel strategy to develop miniature plants with reduced shoot length and aerial coverage but increased branching and floral density. The hairy roots generated from cut *Mecardonia* shoots easily regenerated shoots in the presence of natural oncogenes of *A. rhizogenes* in hormone-free media, and expression of *A. rhizogenes* genes, *root locus D*, *open reading frame 8* (*ORF8*) and *ORF13*, correlated with different observed shoot traits (Pérez de la Torre et al. 2018). This type of approach generates non-GM plants for easy acceptance in commercialisation.

4.7 Genetic Engineering for Stress Tolerance of Ornamental Plants

Much like other cultivated plants, production of ornamentals is severely affected by environmental abiotic stresses like heat, salt, drought and nutrient deficiency, as well as by biotic stress, that is the infestation of pathogens and pests. Genetic engineering has been attempted to alleviate stress tolerance in some ornamental plants. Ectopic expression of orthologs of genes with tested function in model plants has been used to engineer stress tolerance in ornamentals, achieved by *Agrobacterium*-mediated transformation or by direct DNA delivery by gene guns.

4.7.1 Abiotic Stress

Constitutive overexpression of *Salt Overly Sensitive 1* (*SOS1*) encoding a plasma membrane Na^+/K^+ antiporter from *C. crassum* and the TF gene *Inducer of CBF Expression 1* (*ICE1*) from *C. dichrum* in the chrysanthemum cultivar 'Jinba' increased salt, drought and cold tolerance of the transgenic plants (An et al. 2014; Song et al. 2014). The aforementioned genes conferred protection to the transgenic plants via beneficial K^+ ion retention and harmful Na^+ extrusion, osmolyte accumulation and reactive oxygen species management under stress. *Cis*-genic chrysanthemums overexpressing *CmWRKY17* transcriptional repressor from *C. morifolium* increased the sensitivity of transgenic plants to salt stress due to suppression of several stress-related genes including ion transporters and *SOS* pathway genes (Li et al. 2015). The opposite approach, that is silencing of such repressors may prove to be an efficient strategy to enhance salt resistance in ornamentals. Cold-tolerant gerbera plants were engineered by overexpressing Arabidopsis Ca^{2+}/H^+ antiporter

CAX1 (Olsen et al. 2015). Ornamental plant growth and yield are compromised by P-deficient soils and certain P transporters regulate P uptake under conditions of deficiency. A root-expressed *Phosphate transporter 1* (*CmPT1*) gene isolated from *C. morifolium* was overexpressed in the same species with a resultant increase in P accumulation and plant biomass under P-deficiency (Liu et al. 2014). Tissue-specific and stress-inducible promoters isolated from different ornamental plants may prove more suitable than the constitutive CaMV *35S* promoter in genetic engineering for stress tolerance (Smirnova et al. 2015).

4.7.2 Biotic Stress

A modified cystatin transgene from the rice was overexpressed in lily (*Lilium longiflorum* 'Nellie White'), delivered by gene gun, to confer resistance to the root nematode pest *Pratylenchus penetrans* (Vieira et al. 2015). The cystatins up taken by the pests from the GM lilies inhibited nematode digestive proteases preventing their growth. Resistance to a major fungal pathogen *Botrytis cinereal* was achieved in 'Star gazer' cultivar of *Lilium* by overexpression of the rice *RCH10* chitinase gene whose product breaks fungal cell wall chitins (Núñez de Cáceres González et al. 2015). In the latter example, *Agrobacterium*-mediated transformation was used for gene transfer. Biocontrol of a deadly fungal pathogen of gladiolus, *Fusarium oxysporum*, was attempted by biolistic-mediated delivery to gladiolus suspension cells of a gene encoding a synthetic antimicrobial peptide D4E1. This peptide kills the fungus by making an ion-leaking channel in its membrane. The gene was expressed under the constitutive CaMV *35S* promoter and regenerated GM gladioli were more resistant to root infection by *Fusarium* (Kamo et al. 2015). Fungal disease powdery mildew, caused by *Podosphaera* sp., affects many ornamental plants. *Mildew resistance locus 1* (*MLO1*) encodes a membrane transporter necessary for the pathogen to enter the plant. RNA interference-based knockdown of homologs of *MLO1* has been found to increase resistance to powdery mildew in rose (Qiu et al. 2015) and petunia (Jiang et al. 2016). Joint overexpression of *Bacillus thuringiensis cry1Ac* and modified *sarcotoxin IA* gene from the fly *Sarcophagaperegrina* conferred tolerance to both lepidopteran insect pests, for example lepidopteran larvae *Helicoverpa armigera* as well as white rust-causing fungus *Puccinia horiana*, respectively, in *C. morifolium. Sarcotoxin IA* transcripts were more abundant and GM chrysanthemums more pathogen-resistant when attached to Arabidopsis *alcohol dehydrogenase* 5′-untranslated region and *heat shock protein 18.2* gene terminator, the indicating importance of enhancer elements in transgene expression (Shinoyama et al. 2015).

4.8 Genetic Engineering For Post-harvest

Delayed withering and longevity of flowers of ornamental plants is a valued trait for the floral industry. To achieve this, one of the major strategies is the genetic control of the biochemical process of senescence or ageing which is mostly controlled by the phytohormone ethylene. The stimulus for senescence, for example ageing or wounding, leads to enhanced production of ethylene. The amino acid methionine is converted to S-adenosyl-L-methionine (SAM) by the enzyme SAM synthase (SAS). SAM produces 1-aminocyclopropane-1-carboxylic acid (ACC) by ACC synthase (ACS). ACC produces ethylene, catalysed by ACC oxidase (ACO). Ethylene binds to its receptors, for example ETR1 and stops the flow of the signal to CONSTITUTIVE TRIPLE RESPONSE1 (CTR1), which in turn removes the suppression of ETHYLENE INSENSITIVE2 (EIN2). EIN2 is cleaved and transported into the nucleus to activate ETHYLENE INSENSITIVE3 (EIN3) and EIN3-LIKE (EIL) TFs. This activates ETHYLENE RESPONSE FACTORs (ERFs) regulating the expression of several downstream genes, for example *Senescence-Associated Genes* (*SAG*s), causing senescence (Olsen et al. 2015). Genetic manipulation of the aforementioned components of the ethylene signalling pathway to reduce ethylene production or sensitivity has led to delayed senescence of different ornamental flowers. Overexpression of mutated Arabidopsis ethylene receptor *Atert1-1* in petunia led to higher flower lifetime as well as insensitivity to exogenous ethylene, with severe growth defects (Wilkinson et al. 1997) but the use of flower-specific promoter could overcome these defects in kalanchoe (Sanikhani et al. 2008), bell-flower (Sriskandarajah et al. 2007) and geranium (Gehl et al. 2018). Genetic transformation with sense or anti-sense ACS and ACO genes led to lowered ethylene and delayed senescence in different ornamental plants including petunia (Chen et al. 2004; Huang et al. 2007). Overexpression of *PhEIN2* in petunia delayed petal senescence in response to exogenous ethylene, an effect more enhanced when stacked with the *Atetr1-1* mutation (Shibuya et al. 2004). Silencing the petunia ERF TF homeodomain-leucine zipper protein (*PhHD-Zip*) by VIGS led to the suppression of *PhACS* and *PhACO* as well as *PhSAG* transcripts leading to delay in flower senescence. Overexpression of *PhHD-Zip* led to the opposite phenotype, that is accelerated flower senescence (Chang et al. 2014).

External application/enhancing cytokinins indirectly affect senescence by delaying the conversion of ACC to ethylene (Mor et al. 1983). On the other hand, cytokinin degrading enzymes increase during senescence in Petunia (van Doorn and Woltering 2008). Hence, the cytokinin biosynthesis gene *isopentenyl transferase*, driven by senescence-associated promoter P_{SAG12} was used to transform petunia plants which showed a delay in flower senescence as a result of increased cytokinin production in flowers (Chang et al. 2003).

Constitutive expression of genes highly expressed in young tissues and not in senescent tissues helps in delaying senescence. Overexpression of the Arabidopsis gene *FOREVER YOUNG FLOWER*, highly expressed in new flowers in bluebell (*Eustoma grandiflorum*) caused delayed senescence by downregulation of *Ethylene*

Response Factors (Chen et al. 2011). Silencing of genes highly expressed during senescence also proved useful to delay senescence. For example, *MjXB3*, encoding a RING zinc-finger ankyrin repeat protein, highly expressed in senescing flowers of *P. hybrida*, was silenced using VIGS in *P. hybrida*, resulting in 20% increase in flower longevity than wild type (Xu et al. 2007). Complete inhibition of protein synthesis in flowers increases floral longevity. Hence, silencing components of the 26S proteasome or ribosomal subunit genes led to longer-lasting GM petunia flowers (Reid and Jiang 2012).

4.9 Genome Editing of Ornamental Plants

High throughput genotyping platforms have become cost-effective in recent years. This has given way to an explosion of genomic data of cultivated plants in general, ornamentals being no exception. Various genome projects have been undertaken in recent years to sequence several ornamental plant species (Table 4.2). The nuclear genomes of valued ornamentals like a wild rose (Nakamura et al. 2018), cultivated hybrid rose (Qi et al. 2018), sunflower (Badouin et al. 2017), petunia (Bombarely et al. 2016), scarlet sage (Dong et al. 2018), orchid (Zhang et al. 2016b), etc., have been sequenced. In species like chrysanthemums where genome assembly is still incomplete, there are available transcriptomic resources (Wang et al. 2017; Sasaki et al. 2017). Even chloroplast genomes of some ornamentals, for example *Osmanthus* (Wang et al. 2019), lavender (Li et al. 2019), petunia (Wong et al. 2019), hydrangea (Lee et al. 2016) and carnation (Raman and Park 2015) have been sequenced.

The next step is to utilise this genomic information to understand the function of genes which regulate desirable ornamental traits. An important tool for understanding gene function is reverse genetics wherein we study the impact of mutated genes. GE has emerged in recent years as a precise tool for generating targeted mutations in the genome, as opposed to T-DNA or transposon-mediated mutagenesis which are laborious and imprecise (Subburaj et al. 2016). GE relies on specific nucleotide-recognising proteins fused to nucleases which cause double-stranded breaks in the genome. For example, zinc finger nucleases (ZFNs) and transcription activator-like effector nucleases (TALENs) systems. However, the choicest of GE technology is the clustered regularly interspersed short palindromic repeats (CRISPR)/CRISPR-associated protein 9 (Cas9) system due to its high efficiency and cost-effectiveness. This consists of a dimer of Cas9 nuclease and a customisable single guide RNA (sgRNA). The sgRNA is designed complementary to the target genome site to guide the Cas9 dimers to make two double-strand breaks at the desired genome location. The double-strand breaks thus created by these nucleases are repaired by non-homologous end-joining or homologous recombination-based repair mechanisms resulting in insertion and/or deletion mutations. CRISPR/Cas9 is usually achieved by *Agrobacterium*-mediated or microprojectile-based transformation systems, but a fraction of transformed plants is found to be transgene-free that is free from the GE components (Zhang et al. 2016a).

Table 4.2 Sequenced ornamental plant genomes and genome editing in ornamental plants

Type of sequence	Species	References	Examples of genome editing
Chloroplast genome	*Osmanthus cooperi*	Wang et al. (2019)	
Chloroplast genome	*Lavandula dentata*	Li et al. (2019)	
Genome and transcriptome	*Rosa hybrida*	Qi et al. (2018)	
Nuclear genome	*Salvia splendens*	Dong et al. (2018)	
Chloroplast genome	*Petunia hybrida*	Wong et al. (2019)	
Nuclear genome	*Rosa chinensis*	Hibrand Saint-Oyant et al. (2018)	
Nuclear genome	*Helianthus annuus*	Badouin et al. 2017)	
Nuclear genome	*Hibiscus syriacus*	Kim et al. (2017)	
Transcriptome, expressed sequence tags	*Chrysanthemum morifolium*	Sasaki et al. (2017), Wang et al. (2017)	Kishi-Kaboshi et al. (2017)
Nuclear genome	*Rosa multiflora*	Nakamura et al. (2018)	
Nuclear genome	*Petunia sp*	Bombarely et al. (2016)	Subburaj et al. (2016), Zhang et al. (2016a)
Nuclear genome, transcriptome	*Dendrobium catenatum*	Zhang et al. (2016b)	Kui et al. (2017)
Nuclear genome	*Ipomoea nil*	Hoshino et al. (2016)	Watanabe et al. (2017, 2018)
Chloroplast genome	*Hydrangea serrata*	Lee et al. (2016)	
Chloroplast genome	*Dianthus superbus*	Raman and Park (2015)	

The diploid ornamental plant petunia (*P. axillaris* and *P. inflate*) parental genomes have been sequenced (Bombarely et al. 2016). This paved the way for the successful application of GE using CRISPR/Cas9 in *P. hybrid* (Zhang et al. 2016a). Petunia leaf disks were sequentially transformed with Cas9 and sgRNA expression cassettes via *Agrobacterium* infection and genome-edited shoots were regenerated. A targeted deletion in the chlorophyll biosynthesis gene, *phytoene desaturase* (*PDS*) could be efficiently achieved in that study resulting in an albino phenotype in petunia. A high percentage of regenerated shoots (up to 87.5%) showed the mutant phenotype. Petunia protoplasts were also found to be amenable to GE by direct delivery of Cas9 and sgRNA (Subburaj et al. 2016). Other examples of *Agrobacterium*-mediated CRISPR/Cas9 GE in recently sequenced ornamentals include orchid, morning glory and chrysanthemum. Assembly of the draft genome of the Chinese ornamental and medicinal orchid *Dendrobium officinale* (Yan et al. 2015) led to the development of a protocol for GE-based mutation from the protocorm of *D. officinale* wherein five endogenous genes for lignocellulose biosynthesis were targeted (Kui et al. 2017). The assembled genome of Japanese morning glory *Ipomoea nil* (Hoshino et al.

2016) led to GE-based generation of flower colour mutants of this ornamental as well as the model horticultural plant. Knockout mutants of anthocyanin biosynthesis gene *dihydroflavonol-4-reductase-B* (*DFR-B*) (Watanabe et al. 2017) and *carotenoid cleavage dioxygenase 4* (*CCD4*) degrading yellow flower pigment (Watanabe et al. 2018) produced by GE resulted in altered flower colours. Moreover, the progeny of different regenerated plants harbouring different mutant alleles stably inherited the flower colour of the respective parent and were also found to be transgene-free (Watanabe et al. 2017).

Due to polyploidy and large genome size, genome assembly in ornamentals like *Chrysanthemum multifolium* has been incomplete (Sasaki et al. 2017). GE is also exceedingly difficult in such polyploid ornamentals due to difficulty in creating simultaneous mutation events across all genome copies. However, multiple copies of an exogenous gene *yellowish-green fluorescent protein* from *Chiridius poppei* (*CpYGFP*) have been used as targets to develop GE protocols in *C. multifolium* (Kishi-Kaboshi et al. 2017). Decreasing fluorescence of CpYGFP upon introduction of the GE cassettes, subsequent regeneration from calli and continuous culture enabled monitoring of the mutagenesis progress in chrysanthemum.

4.10 Concluding Remarks

Commercialisation of genetically engineered ornamentals is subject to many legal concerns worldwide very much like other GM crops. However, GE using CRISPR/Cas9 will lead to transgene-free non-GM ornamental plants which may be more easily acceptable for commercialisation than GM ornamentals ectopically expressing genes. With the increase in genome sequences available for many commercially important ornamentals, this will soon become routine. These will also pave the way for functional genomics studies in these species which will lead to the identification of genes suitable for manipulating desirable traits. Genome assembly and analysis in some polyploid ornamentals like chrysanthemums and roses are yet to be completed, but transcriptome information and genome of related species will aid in gene identification. Cues from model plants can also help to translate biological information into ornamentals. The concern of GE components remaining in the genome-edited ornamentals can be potentially overcome by direct DNA delivery into protoplasts which has also been demonstrated to be feasible. Protocols for plant regeneration from such protoplasts as well as efficient strategies for transgene-free stable GE have to be developed and/or improved. Hence the future of gene manipulation in ornamentals for achieving desirable traits is bright.

References

An J, Song A, Guan Z et al (2014) The over-expression of *Chrysanthemum crassum CcSOS1* improves the salinity tolerance of chrysanthemum. Mol Biol Rep 41:4155–4162. https://doi.org/10.1007/s11033-014-3287-2

Badouin H, Gouzy J, Grassa CJ et al (2017) The sunflower genome provides insights into oil metabolism, flowering and Asterid evolution. Nature 546:148–152. https://doi.org/10.1038/nature22380

Belarmino MM, Mii M (2000) *Agrobacterium*-mediated genetic transformation of a phalaenopsis orchid. Plant Cell Rep 19:435–442. https://doi.org/10.1007/s002990050752

Bombarely A, Moser M, Amrad A et al (2016) Insight into the evolution of the Solanaceae from the parental genomes of *Petunia hybrida*. Nat Plants. https://doi.org/10.1038/nplants.2016.74

Chang H, Jones ML, Banowetz GM, Clark DG (2003) Overproduction of cytokinins in petunia flowers transformed with P_{SAG12}-IPT delays corolla senescence and decreases sensitivity to ethylene. Plant Physiol 132:2174–2183. https://doi.org/10.1104/pp.103.023945

Chang X, Donnelly L, Sun D et al (2014) A petunia homeodomain-leucine zipper protein, PhHD-zip, plays an important role in flower senescence. PLoS ONE 9:e88320. https://doi.org/10.1371/journal.pone.0088320

Chen J-C, Jiang C-Z, Gookin T et al (2004) Chalcone synthase as a reporter in virus-induced gene silencing studies of flower senescence. Plant Mol Biol 55:521–530. https://doi.org/10.1007/s11103-004-0590-7

Chen MK, Hsu WH, Lee PF, Thiruvengadam M, Chen HI et al (2011) The MADS box gene, *FOREVER YOUNG FLOWER*, acts as a repressor controlling floral organ senescence and abscission in Arabidopsis. Plant J 68:168–185

Chin DP, Shiratori I, Shimizu A et al (2018) Generation of brilliant green fluorescent petunia plants by using a new and potent fluorescent protein transgene. Sci Rep 8:16556. https://doi.org/10.1038/s41598-018-34837-2

Dhanya KG, Thangavel M (2015) Biolistic transformation of rose. Int J Adv Res 3:519–522

Dong A-X, Xin H-B, Li Z-J et al (2018) High-quality assembly of the reference genome for scarlet sage, *Salvia splendens*, an economically important ornamental plant. Gigascience 7. https://doi.org/10.1093/gigascience/giy068

Du L, Lou Q, Zhang X et al (2014) Construction of flower-specific chimeric promoters and analysis of their activities in transgenic *Torenia*. Plant Mol Biol Rep 32:234–245. https://doi.org/10.1007/s11105-013-0646-4

Fu Y, Esselink GD, Visser RGF et al (2016) Transcriptome analysis of gerbera hybrida including in silico confirmation of defense genes found. Front Plant Sci. https://doi.org/10.3389/fpls.2016.00247

Gehl C, Wamhoff D, Schaarschmidt F, Serek M (2018) Improved leaf and flower longevity by expressing the *etr1-1* allele in *Pelargonium zonale* under control of FBP1 and SAG12 promoters. Plant Growth Regul 86:351–363. https://doi.org/10.1007/s10725-018-0434-0

Hibrand Saint-Oyant L, Ruttink T, Hamama L et al (2018) A high-quality genome sequence of *Rosa chinensis* to elucidate ornamental traits. Nat Plants 4(7):473–484. https://doi.org/10.1038/s41477-018-0166-1

Hiratsu K, Matsui K, Koyama T, Ohme-Takagi M (2003) Dominant repression of target genes by chimeric repressors that include the EAR motif, a repression domain, in Arabidopsis. Plant J 34:733–739. https://doi.org/10.1046/j.1365-313X.2003.01759.x

Hoshino A, Jayakumar V, Nitasaka E et al (2016) Genome sequence and analysis of the Japanese morning glory Ipomoea nil. Nat Commun 7:13295. https://doi.org/10.1038/ncomms13295

Hsing H-X, Lin Y-J, Tong C-G et al (2016) Efficient and heritable transformation of *Phalaenopsis* orchids. Bot Stud 57:30. https://doi.org/10.1186/s40529-016-0146-6

Huang L-C, Lai U-L, Yang S-F et al (2007) Delayed flower senescence of *Petunia hybrida* plants transformed with antisense broccoli ACC synthase and ACC oxidase genes. Postharvest Biol Technol 46:47–53. https://doi.org/10.1016/j.postharvbio.2007.03.015

Iantcheva A (2015) Somatic embryogenesis and genetic transformation of carnation (*Dianthus caryophyllus* L.). In: Somatic embryogenesis in ornamentals and its applications. Springer, India, New Delhi, pp 107–120

Jeknić Z, Jeknić S, Jevremović S et al (2014) Alteration of flower color in *Iris germanica* L. 'Fire Bride' through ectopic expression of phytoene synthase gene (*crtB*) from *Pantoea agglomerans*. Plant Cell Rep 33:1307–1321. https://doi.org/10.1007/s00299-014-1617-4

Jiang P, Chen Y, Wilde HD (2016) Reduction of *MLO1* expression in petunia increases resistance to powdery mildew. Sci Hortic (Amsterdam) 201:225–229. https://doi.org/10.1016/j.scienta.2016.02.007

Kamo K (2014) Long term, transgene expression in *Lilium longiflorum* 'Nellie White' grown outdoors and in the greenhouse. Sci Hortic (Amsterdam) 167:158–163. https://doi.org/10.1016/j.scienta.2013.12.011

Kamo K, Lakshman D, Bauchan G et al (2015) Expression of a synthetic antimicrobial peptide, D4E1, in Gladiolus plants for resistance to *Fusarium oxysporum* f. sp. *gladioli*. Plant Cell Tissue Organ Cult 121:459–467. https://doi.org/10.1007/s11240-015-0716-4

Kim Y-M, Kim S, Koo N, et al (2017) Genome analysis of *Hibiscus syriacus* provides insights of polyploidization and indeterminate flowering in woody plants. DNA Res dsw049. https://doi.org/10.1093/dnares/dsw049

Kishi-Kaboshi M, Aida R, Sasaki K (2017) Generation of gene-edited *Chrysanthemum morifolium* using multi-copy transgenes as targets and markers. Plant Cell Physiol pcw222. https://doi.org/10.1093/pcp/pcw222

Kui L, Chen H, Zhang W et al (2017) Building a genetic manipulation tool box for orchid biology: identification of constitutive promoters and application of CRISPR/Cas9 in the orchid *Dendrobium officinale*. Front Plant Sci 7:2036. https://doi.org/10.3389/fpls.2016.02036

Lee J, Lee S-C, Joh HJ et al (2016) The complete chloroplast genome sequence of a Korean indigenous ornamental plant *Hydrangea serrata* for. fertilis Nakai (Hydrangeaceae). Mitochondrial DNA Part B 1:27–28. https://doi.org/10.1080/23802359.2015.1137798

Lee YI, Chen MC, Lin L et al (2018) Increased expression of 9-cis-epoxycarotenoid dioxygenase, *PtNCED1*, associated with inhibited seed germination in a terrestrial orchid, *Phaius tankervilliae*. Front Plant Sci 9. https://doi.org/10.3389/fpls.2018.01043

Liu P, Chen S, Song A et al (2014) A putative high affinity phosphate transporter, CmPT1, enhances tolerance to Pi deficiency of chrysanthemum. BMC Plant Biol 14:18. https://doi.org/10.1186/1471-2229-14-18

Li P, Song A, Gao C et al (2015) Chrysanthemum *WRKY* gene *CmWRKY17* negatively regulates salt stress tolerance in transgenic chrysanthemum and Arabidopsis plants. Plant Cell Rep 34:1365–1378. https://doi.org/10.1007/s00299-015-1793-x

Li H, Li J, Bai H et al (2019) The complete chloroplast genome sequence of *Lavandula dentata* (Lamiaceae) and its phylogenetic analysis. Mitochondrial DNA Part B 4:2135–2136. https://doi.org/10.1080/23802359.2019.1623107

Mor Y, Spiegelstein H, Halevy AH (1983) Inhibition of ethylene biosynthesis in carnation petals by cytokinin. Plant Physiol 71:541–546. https://doi.org/10.1104/pp.71.3.541

Murashige T, Skoog F (1962) A revised medium for rapid growth and bio assays with tobacco tissue cultures. Physiol Plant 15:473–497. https://doi.org/10.1111/j.1399-3054.1962.tb08052.x

Naing AH, Ai TN, Jeon SM et al (2016) An efficient protocol for *Agrobacterium*-mediated genetic transformation of recalcitrant chrysanthemum cultivar Shinma. Acta Physiol Plant 38:38. https://doi.org/10.1007/s11738-015-2059-5

Nakamura N, Katsumoto Y, Brugliera F et al (2015) Flower color modification in *Rosa hybrida* by expressing the *S*-adenosylmethionine: anthocyanin 3′,5′-*O*-methyltransferase gene from *Torenia hybrida*. Plant Biotechnol 32:109–117. https://doi.org/10.5511/plantbiotechnology.15.0205a

Nakamura N, Hirakawa H, Sato S et al (2018) Genome structure of *Rosa multiflora*, a wild ancestor of cultivated roses. DNA Res. https://doi.org/10.1093/dnares/dsx042

Núñez de Cáceres González FF, Davey MR, Cancho Sanchez E, Wilson ZA (2015) Conferred resistance to *Botrytis cinerea* in *Lilium* by overexpression of the *RCH10* chitinase gene. Plant Cell Rep 34:1201–1209. https://doi.org/10.1007/s00299-015-1778-9

Oliva M, Ovadia R, Perl A et al (2015) Enhanced formation of aromatic amino acids increases fragrance without affecting flower longevity or pigmentation in *Petunia* × *hybrida*. Plant Biotechnol J 13:125–136. https://doi.org/10.1111/pbi.12253

Olsen A, Lütken H, Hegelund JN, Müller R (2015) Ethylene resistance in flowering ornamental plants—improvements and future perspectives. Hortic Res 2:15038. https://doi.org/10.1038/hortres.2015.38

Pandey V, Niranjan A, Atri N et al (2014) *WsSGTL1* gene from *Withania somnifera*, modulates glycosylation profile, antioxidant system and confers biotic and salt stress tolerance in transgenic tobacco. Planta 239:1217–1231. https://doi.org/10.1007/s00425-014-2046-x

Pérez de la Torre MC, Fernández P, Greppi JA et al (2018) Transformation of *Mecardonia* (Plantaginaceae) with wild-type *Agrobacterium rhizogenes* efficiently improves compact growth, branching and flower related ornamental traits. Sci Hortic (Amsterdam) 234:300–311. https://doi.org/10.1016/j.scienta.2018.02.047

Ping C-Y, Chen F-C, Cheng T-C et al (2018) Expression profiles of phosphoenolpyruvate carboxylase and phosphoenolpyruvate carboxylase kinase genes in phalaenopsis, implications for regulating the performance of crassulacean acid metabolism. Front Plant Sci 9. https://doi.org/10.3389/fpls.2018.01587

Polturak G, Breitel D, Grossman N et al (2016) Elucidation of the first committed step in betalain biosynthesis enables the heterologous engineering of betalain pigments in plants. New Phytol 210:269–283. https://doi.org/10.1111/nph.13796

Qiu X, Wang Q, Zhang H, Jian H, Zhou N, Ji C, Yan H, Bao M (2015) Antisense *RhMLO1* gene transformation enhances resistance to the powdery mildew pathogen in *Rosa multiflora*. Plant Mol Biol Rep 33:1659–1665

Qi Y, Du L, Quan Y et al (2014) *Agrobacterium*-mediated transformation of embryogenic cell suspension cultures and plant regeneration in *Lilium tenuifolium* oriental × trumpet 'Robina.' Acta Physiol Plant 36:2047–2057. https://doi.org/10.1007/s11738-014-1582-0

Qi W, Chen X, Fang P et al (2018) Genomic and transcriptomic sequencing of *Rosa hybrida* provides microsatellite markers for breeding, flower trait improvement and taxonomy studies. BMC Plant Biol 18:119. https://doi.org/10.1186/s12870-018-1322-5

Raman G, Park S (2015) Analysis of the complete chloroplast genome of a medicinal plant, *Dianthus superbus* var. *longicalyncinus*, from a comparative genomics perspective. PLoS One 10:e0141329. https://doi.org/10.1371/journal.pone.0141329

Reid MS, Jiang C-Z (2012) Postharvest biology and technology of cut flowers and potted plants. Horticultural reviews. Wiley, Hoboken, NJ, USA, pp 1–54

Sahagun J, Kongbangkerd A, Ratanasut K (2018) Organogenic potential of *Dendrobium* floral tissues for stable transformation applications. Philipp J Sci 147:667–676

Sanikhani M, Mibus H, Stummann BM, Serek M (2008) *Kalanchoe blossfeldiana* plants expressing the Arabidopsis *etr1-1* allele show reduced ethylene sensitivity. Plant Cell Rep 27:729–737. https://doi.org/10.1007/s00299-007-0493-6

Sasaki N, Nakayama T (2015) Achievements and perspectives in biochemistry concerning anthocyanin modification for blue flower coloration. Plant Cell Physiol 56:28–40. https://doi.org/10.1093/pcp/pcu097

Sasaki K, Kato K, Mishima H et al (2014) Generation of fluorescent flowers exhibiting strong fluorescence by combination of fluorescent protein from marine plankton and recent genetic tools in *Torenia fournieri* Lind. Plant Biotechnol 31:309–318. https://doi.org/10.5511/plantbiotechnology.14.0907a

Sasaki K, Yamaguchi H, Kasajima I et al (2016) Generation of novel floral traits using a combination of floral organ-specific promoters and a chimeric repressor in *Torenia fournieri* lind. Plant Cell Physiol 57:1319–1331. https://doi.org/10.1093/pcp/pcw081

Sasaki K, Mitsuda N, Nashima K et al (2017) Generation of expressed sequence tags for discovery of genes responsible for floral traits of *Chrysanthemum morifolium* by next-generation sequencing technology. BMC Genomics 18:683. https://doi.org/10.1186/s12864-017-4061-3

Schwinn KE, Boase MR, Bradley JM et al (2014) MYB and bHLH transcription factor transgenes increase anthocyanin pigmentation in petunia and lisianthus plants, and the petunia phenotypes are strongly enhanced under field conditions. Front Plant Sci 5:603. https://doi.org/10.3389/fpls.2014.00603

Shen Y, Xing W, Ding M et al (2016) Somatic embryogenesis and *Agrobacterium*-mediated genetic transformation in rosa species. Somatic embryogenesis in ornamentals and its applications. Springer India, New Delhi, pp 169–185

Shibuya K, Barry KG, Ciardi JA et al (2004) The central role of PhEIN2 in ethylene responses throughout plant development in petunia. Plant Physiol 136:2900–2912. https://doi.org/10.1104/pp.104.046979

Shinoyama H, Sano T, Saito M et al (2012) Induction of male sterility in transgenic chrysanthemums (*Chrysanthemum morifolium* Ramat.) by expression of a mutated ethylene receptor gene, *Cm-ETR1/H69A*, and the stability of this sterility at varying growth temperatures. Mol Breed 29:285–295. https://doi.org/10.1007/s11032-010-9546-6

Shinoyama H, Mitsuhara I, Ichikawa H et al (2015) Transgenic chrysanthemums (*Chrysanthemum morifolium* Ramat.) carrying both insect and disease resistance. Acta Hortic 485–497. https://doi.org/10.17660/ActaHortic.2015.1087.66

Smirnova OG, Tishchenko EN, Ermakov AA et al (2015) Promoters for transgenic horticultural plants. Abiotic stress biology in horticultural plants. Springer Japan, Tokyo, pp 169–186

Song A, An J, Guan Z et al (2014) The constitutive expression of a two transgene construct enhances the abiotic stress tolerance of chrysanthemum. Plant Physiol Biochem 80:114–120. https://doi.org/10.1016/j.plaphy.2014.03.030

Sriskandarajah S, Mibus H, Serek M (2007) Transgenic *Campanula carpatica* plants with reduced ethylene sensitivity. Plant Cell Rep 26:805–813. https://doi.org/10.1007/s00299-006-0291-6

Subburaj S, Chung SJ, Lee C et al (2016) Site-directed mutagenesis in Petunia × hybrida protoplast system using direct delivery of purified recombinant Cas9 ribonucleoproteins. Plant Cell Rep 35:1535–1544. https://doi.org/10.1007/s00299-016-1937-7

Tanaka Y, Brugliera F von MC (2013) Flower colour and cytochromes P450. Philos Trans R Soc Lond B Biol Sci 368:20120432

Teixeira da Silva JA, Dobránszki J, Cardoso JC et al (2016) Methods for genetic transformation in Dendrobium. Plant Cell Rep 35:483–504. https://doi.org/10.1007/s00299-015-1917-3

Unver T, Budak H (2009) Virus-induced gene silencing, a post transcriptional gene silencing method. Int J Plant Genomics 2009:198680

van Doorn WG, Woltering EJ (2008) Physiology and molecular biology of petal senescence. J Exp Bot 59:453–480. https://doi.org/10.1093/jxb/erm356

Vieira P, Wantoch S, Lilley CJ et al (2015) Expression of a cystatin transgene can confer resistance to root lesion nematodes in *Lilium longiflorum* cv. 'Nellie White.' Transgenic Res 24:421–432. https://doi.org/10.1007/s11248-014-9848-2

Villacorta-Martin C, Núñez de Cáceres González FF, de Haan J et al (2015) Whole transcriptome profiling of the vernalization process in *Lilium longiflorum* (cultivar White Heaven) bulbs. BMC Genomics 16:550. https://doi.org/10.1186/s12864-015-1675-1

Wang J, Wang H, Ding L et al (2017) Transcriptomic and hormone analyses reveal mechanisms underlying petal elongation in *Chrysanthemum morifolium* 'Jinba.' Plant Mol Biol 93:593–606. https://doi.org/10.1007/s11103-017-0584-x

Wang X, Cai F, Zhang C et al (2019) Characterization of the complete chloroplast genome of the ornamental plant *Osmanthus cooperi*. Mitochondrial DNA Part B 4:2314–2315. https://doi.org/10.1080/23802359.2019.1627951

Watanabe K, Kobayashi A, Endo M et al (2017) CRISPR/Cas9-mediated mutagenesis of the dihydroflavonol-4-reductase-B (DFR-B) locus in the Japanese morning glory *Ipomoea (Pharbitis) nil*. Sci Rep 7:10028. https://doi.org/10.1038/s41598-017-10715-1

Watanabe K, Oda-Yamamizo C, Sage-Ono K et al (2018) Alteration of flower colour in Ipomoea nil through CRISPR/Cas9-mediated mutagenesis of carotenoid cleavage dioxygenase 4. Transgenic Res 27:25–38. https://doi.org/10.1007/s11248-017-0051-0

Wen C, Zhao Q, Nie J et al (2016) Physiological controls of chrysanthemum *DgD27* gene expression in regulation of shoot branching. Plant Cell Rep 35:1053–1070. https://doi.org/10.1007/s00299-016-1938-6

Wilkinson JQ, Lanahan MB, Clark DG et al (1997) A dominant mutant receptor from Arabidopsis confers ethylene insensitivity in heterologous plants. Nat Biotechnol 15:444–447. https://doi.org/10.1038/nbt0597-444

Wong J, Mudd EA, Hayes A, Day A (2019) The chloroplast genome sequence of the ornamental plant *Petunia hybrida*. Mitochondrial DNA Part B 4:249–250. https://doi.org/10.1080/23802359.2018.1547136

Wu J, Liu CC, Seng S et al (2014) Somatic embryogenesis and *Agrobacterium*-mediated transformation of *Gladiolus hybridus* cv. 'Advance Red.' Plant Cell Tissue Organ Cult 120:717–728. https://doi.org/10.1007/s11240-014-0639-5

Xie Q, Chen G, Liu Q et al (2015) Dual silencing of *DmCPD* and *DmGA20ox* genes generates a novel miniature and delayed-flowering *Dendranthema morifolium* variety. Mol Breed 35:67. https://doi.org/10.1007/s11032-015-0239-z

Xu X, Jiang C-Z, Donnelly L, Reid MS (2007) Functional analysis of a RING domain ankyrin repeat protein that is highly expressed during flower senescence. J Exp Bot 58:3623–3630. https://doi.org/10.1093/jxb/erm212

Yan L, Wang X, Liu H et al (2015) The genome of *Dendrobiumofficinale* Illuminates the biology of the important traditional chinese orchid herb. Mol Plant 8:922–934. https://doi.org/10.1016/j.molp.2014.12.011

Yue Y, Yu R, Fan Y (2015) Transcriptome profiling provides new insights into the formation of floral scent in *Hedychium coronarium*. BMC Genomics 16:470. https://doi.org/10.1186/s12864-015-1653-7

Zhang B, Yang X, Yang C et al (2016a) Exploiting the CRISPR/Cas9 system for targeted genome mutagenesis in petunia. Sci Rep 6:20315. https://doi.org/10.1038/srep20315

Zhang GQ, Xu Q, Bian C et al (2016b) The *Dendrobium catenatum* Lindl. genome sequence provides insights into polysaccharide synthase, floral development and adaptive evolution. Sci Rep. https://doi.org/10.1038/srep19029

Zhao D, Shi Y, Senthilkumar HA et al (2019) Enriched networks 'nucleoside/nucleotide and ribonucleoside/ribonucleotide metabolic processes' and 'response to stimulus' potentially conferred to drought adaptation of the epiphytic orchid *Dendrobium wangliangii*. Physiol Mol Biol Plants 25:31–45. https://doi.org/10.1007/s12298-018-0607-3

Zheng X, Liu S, Cheng C et al (2016) Cloning and expression analysis of betalain biosynthesis genes in *Amaranthus tricolor*. Biotechnol Lett 38:723–729. https://doi.org/10.1007/s10529-015-2021-z

Zhong X, Yuan X, Wu Z et al (2014) Virus-induced gene silencing for comparative functional studies in *Gladiolus hybridus*. Plant Cell Rep 33:301–312. https://doi.org/10.1007/s00299-013-1530-2

Chapter 5
Enhancing Abiotic Stress Tolerance in Plants Through Genome Editing

Deepa Jaganathan, Rohit Kambale, Hifzur Rahman, Devanand Pachanoor Subbian, and Raveendran Muthurajan

Abstract The ever-increasing population and predicted climate change demands the use of technological advancements inbreeding to achieve a sustained increase in crop productivity. Availability of genome sequence information of major crops enabled the development of genomics-assisted breeding approaches aimed at assembling trait-specific genes using innovative mating designs. Further advancements in genome engineering enabled accelerated trait improvement using CRISPR/Cas9 mediated targeted mutagenesis. Creation and selection of superior alleles through genome editing approaches is one of the potential applications of genome engineering. Advancement made to the traditional CRISPR/Cas9 including identification and modification of Cas9 variants, base editing and multiplex editing paved the way towards precision genome engineering. Sustained efforts in this direction may lead to the development of improved crop varieties to combat multiple abiotic stresses. In this review, applications of CRISPR/Cas9 tools in the genetic manipulation of abiotic stress tolerance in crops are discussed.

Keywords Abiotic stress tolerance · Genome engineering · CRISPR/Cas9

D. Jaganathan · R. Kambale · H. Rahman · D. P. Subbian · R. Muthurajan (✉)
Centre for Plant Molecular Biology and Biotechnology, Tamil Nadu Agricultural University, Coimbatore 641003, India
e-mail: raveendrantnau@gmail.com

D. Jaganathan
e-mail: deebiotech@gmail.com

R. Kambale
e-mail: rohitkamble568@gmail.com

H. Rahman
e-mail: rahman.biotech@gmail.com

D. P. Subbian
e-mail: devps7@yahoo.com

A. Bhattacharya et al. (eds.), *CRISPR/Cas Genome Editing*, Concepts and Strategies in Plant Sciences, https://doi.org/10.1007/978-3-030-42022-2_5

5.1 Introduction

The global population is expected to cross 9 billion by 2050 indicating the need for increasing food production by 70–100% (Tilman et al. 2011). Achieving this target needs innovations in crop breeding technology to achieve a sustained increase in agricultural productivity. Current productivity growth in major crops will not be sufficient to meet the predicted demand (Ray et al. 2013). In addition, climate change (Thornton 2012) poses a further threat to agriculture by increasing the frequency of occurrence of drought, salinity, submergence, temperature extremes etc. Hybridisation, mutation breeding and transgenic breeding are the major breeding tools in modern agriculture and they take several years to introduce desirable alleles by breeding and to increase variability by genetic recombination. In addition, transgenic breeding is not accepted by people in several countries.

The success of any crop breeding depends mainly on the genetic variation within a species that provide more opportunities to find and combine desirable characteristics. To enhance variability in the native gene pool, plant breeders started using mutation breeding to create new crop traits. Mutation breeding has led to the development of >3000 improved crop varieties in more than 175 crops including rice, maize, wheat, banana, tomato, pumpkin and soya. Crops obtained via mutation breeding have been safely cultivated and eaten for decades. Major drawbacks with mutation breeding are randomness of the mutation in the genome and low frequency of desirable traits. This warrants development of new tools for accelerated development of abiotic stress-tolerant genotypes in major crops.

Genome editing technology has become an effective tool for improving plant traits by causing desired changes in the target genes. Targeted alteration of genes in crop improvement uses five different tools viz., (1) Oligonucleotide Directed Mutagenesis (ODM); (2) Zinc Finger Nucleases; (3) Meganucleases; (4) Transcription Activator-Like Effector nucleases (TALENs) and (5) Clustered Regularly Interspaced Short Palindromic Repeats—Associated Systems (CRISPR/Cas9). Among the five different systems, TALENs and CRISPR/Cas9 systems have become popular and widely used in plants. Clustered regularly interspaced palindromic repeats (CRISPR) or CRISPR/Cas9 was reported as the bacterial adaptive immune system and adopted for genetic modification of animal and plant genomes (Jinek et al. 2012; Jiang et al. 2013). In the past five years enormous studies have been reporting the utilisation of CRISPR/Cas9 based genome editing approach in crops (Bortesi and Fischer 2015; Song et al. 2016; Jaganathan et al. 2018; Zhang et al. 2019a, b; Li et al. 2019a, b). Since its first demonstration for genome editing, several modifications and improvements has been made in CRISPR/Cas9 genome editing. Cas9 is a type II CRISPR system from *Streptococcus pyogenes* (SpCas9) which is adopted widely. However, several alternatives are being reported from different species for precise genome editing (Zhang et al. 2019b). Various types of Cas proteins allows a wide range of genome modifying applications (Fig. 5.1). It includes, (i) CRISPR/Cas9, which makes a blunt end dsDNA cut and repaired through nonhomilogoues end joining (NHEJ) or homologous end joining with a donor DNA template. It is majorly adopted

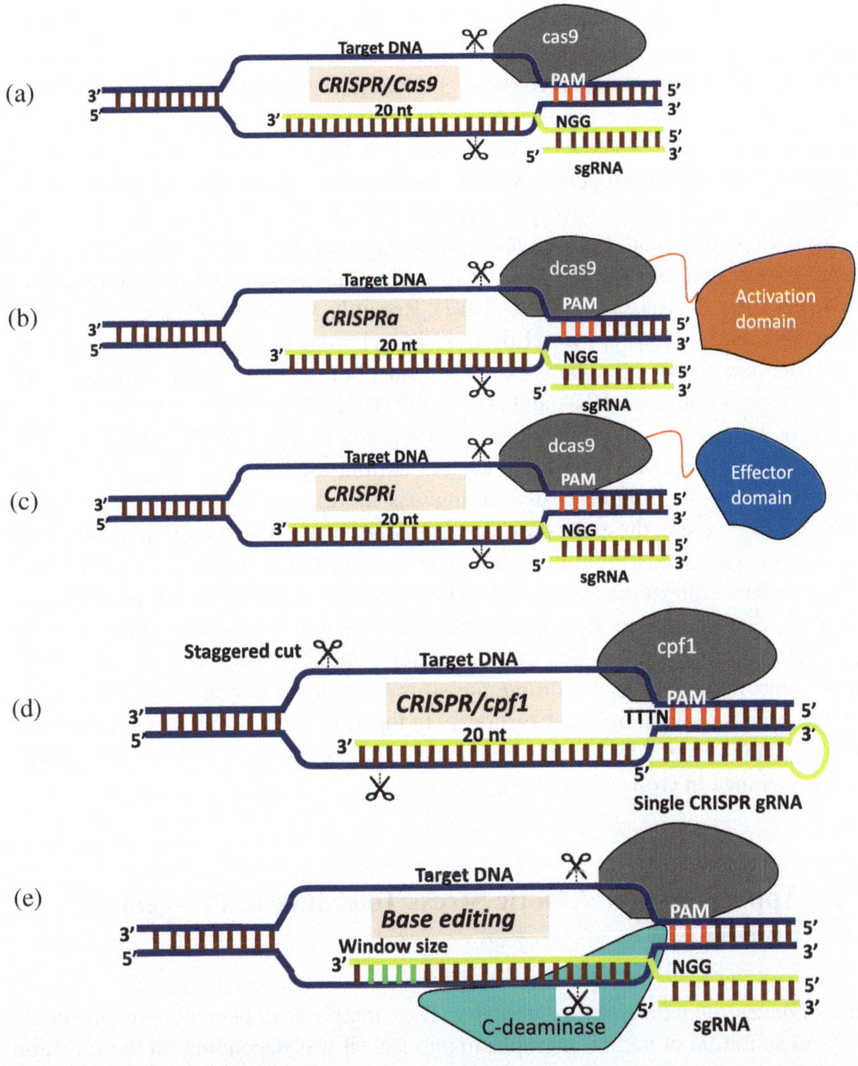

Fig. 5.1 Various CRISPR tools for genome editing. a CRISPR/Cas9 is a traditional gene-editing approach which is currently the most widely used method for loss of function studies. **b** Catalytically inactive version of cas9 is used for enhanced expression of the target genes by targeting transcriptional activators. **c** Fusion of repressor domains to dcas9 allows gene knockdown instead of knockouts. **d** Cpf1 is preferred over Cas9 due to its staggered cut and 'T' rich PAM region. **e** Base editors such as adenine base editor and cytidine base editors are used for precise cleavage at SNP sites

for knock out experiments where the target gene is a negative regulator. Cas9 recognises 3′-NGG PAM site (Jinek et al. 2012), whereas its orthologues such as saCas9 from *Staphylococcus aureus* recognises 3′-NNGRRT (Ran et al. 2015) FnCas9 from *Francisella novicida* recognises 5′-NGG PAM and NmeCas9 from *Neisseria meningitidis* recognises 3′-NNNNGATT (Lee et al. 2016) (ii) CRISPR/Cas12, a type V system which makes staggered dsDNA cut and it recognises T-rich PAM sites (5′-TTTN). It possesses two subtypes Type V-A (Cpf1) and Type V-B (C2c1). Cpf1 is from *Prevotella* and *Francisella* 1, it allows multiple rounds of cutting at the target region as it cuts the DNA distal from the PAM region and it requires a short CRISPR RNA (CrRNA) (Mahfouz 2017). Recently, Cpf1 variants such as AsCpf1 from *Acidaminococcus* sp. BV3L6) and LbCpf1 from *Lachnospiraceae* bacterium ND2006 were reported for plant genome editing (Tang et al. 2017). In contrast to cpf1, C2c1 requires both tracr RNA and crRNA (iii) CRISPR/Cas13 also known as C2c2 is a Type VI system and unlike other, it targets single-strand RNA instead of DNA. In addition, a promising tool for precisely editing the causative single nucleotide polymorphism (SNP) namely base editing was also discovered in recent years. Base editing allows to edit the exact base which renders single base mutation by altering a cytidine nucleotide to a thymine nucleotide (termed Cytidine base editor, CBE, alters C to T in the coding strand or G to A in the negative strand (Komor et al. 2016) and an adenine nucleotide to a guanine nucleotide (Adenine base editor, ABE, alters A to G in the coding strand or T to C in the negative strand; (Gaudelli et al. 2017). With these improvements in gene editing, now it is possible for targeting a wide range of genes involved in various stress response. In this review, we summarise and discuss the applications of CRISPR/Cas9 genome editing approaches for improving abiotic stress tolerance in crops (Table 5.1).

5.2 Approaches for Abiotic Stress Tolerance in Pre-genome Editing Era

(i) *Conventional breeding approaches*: The early phase of plant breeding involved exploitation of natural germplasm and later it was depending on the selection for component traits like higher yield and short duration. Introduction of semi-dwarf rice and wheat varieties during the first green revolution had eliminated several stress-tolerant genotypes. Due to the genome complexity, backcross breeding procedures and gene pyramiding took several years for developing a stress-tolerant variety.

(ii) *Marker-assisted selection*: Advancements in molecular genetics discovered genetic loci linked to various quantitative traits in major crop plants which enabled the deployment of marker-assisted selection (MAS) in crop improvement. MAS is efficient over classical breeding as it eases the selection during the early stage of crop growth and therefore it reduced the time and cost (Varshney et al. 2005). MAS can be used efficiently for pyramiding multiple genes in a

Table 5.1 List of genes targeted for abiotic stress tolerance using transgenic approaches

Crop	Gene/QTLs	Symbol (Gene/QTL)	Type	Target trait	References
Arabidopsis	*Miscanthuslutarioriparius NAC gene*	*MINAC5*	Overexpression	Drought, cold	Yang et al. (2015)
Arabidopsis	*Dehydration-responsive element-binding*	*GmDREB1s*	Overexpression	Drought, cold, salt, heat	Kidokoro et al. (2015)
Arabidopsis	*NAM, ATAF and CUC (NAC) transcription factors*	*NAC*	Overexpression	Drought, salt	Liu et al. (2011)
Arabidopsis	*Arabidopsis thaliana Zeaxanthin epoxidase*	*AtZEP*	Overexpression	Drought	Schwartz et al. (2003)
Arabidopsis	*Dehydration-responsive element-binding (DREB)*	*DREB2A*	Overexpression	Drought	Sakuma et al. (2006)
Rice	*Tocopherol cyclase*	*Tocopherol cyclase*	Overexpression	Drought	Woo et al. (2014)
Rice	*Pyrabactin resistance*	*PYL5*	Overexpression	Drought	Kim et al. (2014)
Rice	*Oryza sativa Abscisic acid 8′-hydroxylase 3*	*OsABA8ox3*	RNAi and overexpression	Drought	Cai et al. (2015)
Rice	*codA gene for choline oxidase*	*CodA*	Overexpression	Salt	Mohanty et al. (2002)
Rice	*Chloroplastic glutamine synthetase*	*GS2*	Overexpression	Salt, cold	Hoshida et al. (2000)
Rice	*Δ1-pyrroline-5-carboxylate synthetase*	*P5CS*	Overexpression	Drought, salt	Zhu et al. (1998)
Rice	*Calcium-dependent protein kinase*	*OSCDPK7*	Overexpression	Cold and salt/drought	Saijo et al. (2000)
Rice	*Late embryogenesis abundant (LEA) protein gene*	*HVA1*	Overexpression	Drought, salt	Xu and Mackill (1996)

(continued)

Table 5.1 (continued)

Crop	Gene/QTLs	Symbol (Gene/QTL)	Type	Target trait	References
Rice	*Late embryogenesis abundant (LEA) protein gene*	*HVA1*	Overexpression	Drought, salt	Rohila et al. (2002)
Rice	*LEA group 2 protein* *LEA group 1 protein*	*PMA80 & PMA1959*	Overexpression	Drought, salt	Cheng et al. (2002)
Rice	*Arginine decarboxylase*	*ADC*	Overexpression	Drought, salt	Roy and Wu (2001)
Rice	*S-adenosylmethionine decarboxylase*	*SAMDC*	Overexpression	Salt	Roy and Wu (2002)
Rice	*Catalase*	*Catalase*	Overexpression	Cold	Matsumura et al. (2002)
Rice	*Trehalose-6-phosphate synthase and trehalose-6-phosphate phosphatase*	*OtsA + OtsB*	Overexpression	Drought, salt, cold	Garg et al. (2002)
Rice	*Trehalose-6-phosphate (T-6-P) synthase and T-6-P phosphatase (TPP)*	*TPS + TPP*	Overexpression	Drought, salt, cold	Jang et al. (2003)
Rice	*Stress-induced transcription factor NAC1*	*SNAC1*	Overexpression	Drought	Hu et al. (2006)
Rice	*ZINC FINGER PROTEIN 252*	*ZFP252*	Overexpression	Drought	Xu et al. (2008)
Tobacco	*Heat Shock protein*	*NtHSP70-1*	Overexpression	Drought	Cho and Hong (2006)
Wheat	*Late embryogenesis abundant (LEA) protein gene*	*HVA1*	Overexpression	Drought	Sivamani et al. (2000)
Wheat	*HIGH-AFFINITY K + TRANSPORTER 1*	*HKT1*	RNAi technology	Salt	Laurie et al. (2002)

(continued)

Table 5.1 (continued)

Crop	Gene/QTLs	Symbol (Gene/QTL)	Type	Target trait	References
Wheat	*Mannitol-1-phosphate dehydrogenase*	*MtlD*	Overexpression	Drought, salt	Abebe et al. (2003)
Mustard	*Glutathione synthetase*	*Glutathione synthetase*	Overexpression	Heavy metal	Zhu et al. (1999)
Mustard	*Glutathione reductase*	*Glutathione reductase*	Overexpression	Heavy metal	Pilon-Smits et al. (2000)
Mustard	*Na+/H+ antiporter*	*AtNHX1*	Overexpression	Salt	Zhang et al. (2001)
Mustard	*Osmotin*	*Osmotin*	Overexpression	Drought, salt	Tayal et al. (2003)
Alfalfa	*Manganese superoxide dismutase (MnSOD)*	*MnSOD*	Overexpression	Cold	McKersie et al. (1993)
Alfalfa	*Manganese superoxide dismutase (MnSOD)*	*MnSOD*	Overexpression	Cold	McKersie et al. (1996)
Alfalfa	*Manganese superoxide dismutase (MnSOD)*	*MnSOD*	Overexpression	Cold	McKersie et al. (1999)
Alfalfa	*Alfin*	*Alfin*	Overexpression	Salt	Winicov and Bastola (1999)
Cotton	*Manganese superoxide dismutase (MnSOD*	*MnSOD*	Overexpression	Cold	Allen (1995)
Cotton	*ascorbate peroxidase gene*	*Apx*	Overexpression	Cold	Payton et al. (2001)
Potato	*Copper- and zinc-containing superoxide dismutase (Cu/ZnSOD)*	*Cu, ZnSOD*	Overexpression	Oxidative stress	Perl et al. (1993)
Potato	*Osmotin like protein*	*Osmotin like protein*	Antisense Technology	Cold	Zhu et al. (1996)
Potato	*CDSP32*	*Osmotin*	Overexpression	Drought, salt	Babu and Bansal (1998)
Potato	*Antifreeze proteins (AFPs)*	*AFP*	Overexpression	Cold	Wallis et al. (1997)

(continued)

Table 5.1 (continued)

Crop	Gene/QTLs	Symbol (Gene/QTL)	Type	Target trait	References
Potato	*Trehalose-6-phosphate synthase*	*TPS1*	Overexpression	Drought	Yeo et al. (2000)
Potato	*Glyceraldehyde-3 phosphate dehydrogenase*	*Glyceraldehyde-3 phosphate dehydrogenase*	Overexpression	Salt	Jeong et al. (2001)
Tomato	*C-repeat-binding factor*	*CBF1*	Overexpression	Salt, drought, oxidative stress	Zhang et al. (2001)
Tomato	*C-repeat-binding factor*	*CBF3*	Overexpression	Cold, oxidative stress	Hsieh et al. (2002)
Oat	*Late embryogenesis abundant (LEA) protein gene*	*HVA1*	Overexpression	Osmotic tolerance	Maqbool et al. (2002)

single cultivar with a high level of precision. Several successful examples have proven utilisation of molecular markers for crop improvement few of them are abiotic stress tolerance in rice (Rahman et al. 2018; Valarmathi et al. 2019) biotic stress tolerance (Chukwu et al. 2019). However, molecular breeding had partially failed to reduce the breeding timeline for crop varietal improvement, as elimination of linkage drag needed several backcrosses. The success of MAS depends on the availability of precise markers closely linked to the trait of interest. Very limited progress was made in QTL mapping and such identified QTLs ranged several centiMorgon (cM) on a genetic map covering huge regions on physical map (Jaganathan et al. 2020). Therefore, introgressing such large QTLs was challenging through MAS. Identifying QTLs with tightly linked markers or located within the gene of interest demands high throughput genotyping and phenotyping. Advancements in phenomics and genomics-assisted molecular breeding techniques are promising tools for crop improvement, however, high cost involved in these techniques limits the implementation of these techniques (Varshney et al. 2005). In rice, SSR markers were widely used for identifying yield QTLs under drought and introgression program (Kumar et al. 2014). Several drought-tolerant QTLs were reported in maize using SNP markers in nested assisted mapping population (Li et al. 2016). In chickpea, SSR markers along with DaRT, CAPS markers were used for the identification of a promising QTL for drought tolerance and was further fine mapped using SNP markers which were successfully introgressed and two superior drought-tolerant chickpea varieties were released (Varshney et al. 2014; Jaganathan et al. 2015, https://icar.org.in/content/development-two-superior-chickpea-varieties-genomics-assisted-breeding). Similarly, the success of genomics-assisted breeding for abiotic and biotic stress tolerance has been evidenced in other legumes including groundnut and pigeonpea (Varshney et al. 2019). A list of genes/QTLs targeted for abiotic stress tolerance through marker-assisted selection is listed in Table 5.2.

(iii) *Transgenics*: Genetic engineering has been proved to be a reliable tool for introducing novel traits by overcoming sexual barriers. Implementation of genetically modified crop plants may reduce the use of pesticides, can improve yield and hence raise farmer's income. Major concerns are time and the cost needed to develop and obtain approval for food, feed as well as research purpose (Zhang et al. 2016).

5.3 Genome Engineering for Drought Stress Tolerance

Drought is a serious limitation to rice production and rice is extremely sensitive to drought (Lafitte et al. 2004). Rainfed areas are subjected to drought frequently at any stage of its growth resulting in reduced crop yields (Babu et al. 2003). However, progress in genetic improvement of crops for drought resistance using conventional approaches has been slow. QTLs for drought tolerance have been mapped in major

Table 5.2 List of genes/QTLs targeted for abiotic stress tolerance through marker-assisted selection

Crop	Gene	Symbol (gene/QTL)	Target trait	References
Rice	*Salinity tolerance*	*Saltol*	Salinity	Thomson et al. (2010)
Rice	*Streptomyces lividans K + Channel*	*SKC1*	Salinity stress	Emon et al. (2015)
Rice	*drought and salt tolerance*	*DST*	Salinity stress	Emon et al. (2015)
Rice	*High-affinity K + Transporter*	*HKT1; 5*	Saline Soil	Ren et al. (2005)
Rice	*DEEPER ROOTING 1*	*DRO1*	Drought	Uga et al. (2011, 2013)
Rice	*QTL for Grain yield under reproductive-stage drought stress*	*qDTY1.1, qDTY2.1*	Drought tolerance	Venuprasad et al. (2009, 2012)
Rice	*QTL for drought*	*qSDT2.1 and qSDT12-2*	Drought tolerance	Zhang et al. (2006)
Rice	*Ethylene response factors SNORKEL1 and SNORKEL2*	*SK1 and SK2*	Flooding	Hattori et al. (2009)
Rice	*SUBMERGENCE 1 (SUB1)*	*Sub1*	Submergence	Xu and Mackill (1996), Neeraja et al. (2008)
Rice	*Nramp aluminium transporter (NRAT1)*	*NRAT1*	High Al^{3+}	Li et al. (2014)
Rice	*Seedling stage chilling tolerance (qSCT-11)*	*qSCT-11*	Chilling-tolerant	Chen and Li (2005)
Rice	*QTL percent seed set in cold water-treated*	*qPSST-3, qPSST-7, and qPSST-9*	Cold stress	Jena et al. (2010)
Rice	*spikelet fertility under heat stress*	*qHTSF4.1*	Heat stress	Ye et al. (2015)
Rice	*Phosphorus uptake1 (Pup1)*	*Pup1*	Phosphorus Use Efficiency	Chin et al. (2010), Gamuyao et al. (2012)
Wheat	*high-affinity potassium transporter1*	*HKT1;5-A at the Nax2 locus*	Saline Soil	Munns et al. (2012), James et al. (2012)
Wheat	*Vernalisation gene*	*VRN1 at the FR1 locus*	Low temperature	Dhillon et al. (2010), Knox et al. (2010)

(continued)

Table 5.2 (continued)

Crop	Gene	Symbol (gene/QTL)	Target trait	References
Wheat	*C-Repeat Binding Factor*	*CBFs at the FR2 locus*	Low temperature	Knox et al. (2010), Skinner et al. (2006), Francia et al. (2004, 2007)
Wheat	*NAC transcription factor (NAM-B1)*	*NAM-B1*	Low Fe^{3+} and Zn^{2+}	Uauy et al. (2006)
Common Wheat		*ALMT*	High Al^{3+}	Sasaki et al. (2004; 2006);
Barley	*Heat-shock protein*	*HSP116.8*	Drought tolerance	Xie and Yang (2013)
Sorghum	*Sorghum bicolor multidrug and toxic compound extrusion Al tolerance locus*	*SbMATE at the AltSB locus*	High Al^{3+}	Magalhaes et al. (2007)
Maize	*Al tolerance gene multidrug and toxic compound extrusion 1 (MATE1)*	*MATE1*	High Al^{3+}	Maron et al. (2013)
Pigeon Pea	*Cyclophilin*	*CcCYP*	Drought, salinity and cold	Priyanka et al. (2010)
Pigeon Pea	*Cold and drought regulatory (CcCDR) genes*	*CcCDR*	Drought, salinity and cold	Pazhamala et al. (2015)
Pigeon Pea	*Cajanuscajan MULTIDRUGS AND TOXIC COMPOUNDS EXCLUSION (CcMATE1)*	*CcMATE1*	Al tolerance	Daspute et al. (2018)
Chickpea	*Abscisic acid stress and ripening (ASR); dehydrin (DHN); dehydration-responsive element-binding (DREB)*	*ASR, DHN, and DREB*	Drought and heat tolerance	Thudi et al. (2014)
Chickpea	*Quantitative trait loci early flowering (QTL/Efl-1)*	*QTL/Efl-1 locus*	Drought escape (flowering time)	Cho et al. (2002)
Chickpea	*Quantitative trait loci photoperiod (QTL/ppd)*	*QTL/ppd*	Drought escape (flowering time)	Lichtenzveig et al. (2006)
Chickpea	*abscisic acid stress and ripening gene (ASR); dehydrin (DHN); dehydration-responsive element-binding (DREB)*	*ASR, DHN, and DREB*	Drought and heat tolerance	Thudi et al. (2014)

crops like rice (Kamoshita et al. 2008; Khowaja et al. 2009; Swamy et al. 2011; Suji et al. 2012), wheat (Kirigwi et al. 2007; Mathews et al. 2008; Pinto et al. 2010; Gahlaut et al. 2017), maize (Hao et al. 2010; Almeida et al. 2014; Trachsel et al. 2016), Chick pea (Rehman et al. 2011; Varshney et al. 2014; Sivasakthi et al. 2018), etc.

CRISPR/Cas9 approach is applied for improving drought tolerance in major crops (Table 5.3). For instance, Lou et al. (2017) characterised the role of *OsSAPK2* in rice by creating a loss of function mutants through CRISPR/Cas9 approach. SNF 1-RELATED PROTEIN KINASE 2 (SnRK2) is a key regulator of hyper-osmotic stress signalling and abscisic acid (ABA)-dependent development in various plants. This study targeted 3rd exon of *OsSAPK2* for sgRNA designing. The mutant lines (*sapk2*) were more sensitive to drought and reactive oxygen species (ROS) than wild-type plants. Phenotypic and expression analysis of the edited plants showed a better survival rate and higher gene expression under drought stress than control plants. CRISPR/Cas9-mediated gene editing of mitogen-activated protein kinases (MAPKs) in tomato (*SlMAPK3*) showed that *SlMAPK3* was involved in drought stress response in tomato (Wang et al. 2017). *slmapk3* mutant lines were generated by targeting 3rd exon of *SlMAPK3* and the mutant lines showed more wilting, higher hydrogen peroxide content, lower antioxidant enzymes activities, and suffered more membrane damage under drought stress. Further, this study concluded that *SlMAPK3* modulates drought responses in tomato by protecting cell membranes from *oxidative damage and modulating transcription of stress-related genes.* Abscisic acid (ABA)-responsive element-binding protein 1/ABRE binding factor (AREB1/ABF2) in Arabidopsis was reported to positively regulate drought tolerance. Paixão et al. (2019) employed CRISPR activation (CRISPRa) system to activate the promoter of AREB1 using CRISPRa $dcas9_{HAT}$. Recently Li et al. (2019a, b) demonstrated the role of pathogenesis-related gene1 (*NPR1*) in drought sensitivity in tomato using CRISPR/cas9 approach. This study confirmed the reduced drought tolerance in the *slnpr1* mutant lines by gene editing of two targets in the first and second exon of *SLNPR1*. Characterisation of *slnpr1* in tomato will further enhance the understanding of drought tolerance mechanism in tomato. Chen et al. (2019) employed CRISPR/Cas9-mediated gene editing approach for improving seed size under abiotic stress conditions (drought and salinity). Two transcription repressors, DPA4 (Development-Related PcG Target in the APEX4)/NGAL3 and SOD7 (Suppressor of da1-1)/NGAL2 (NGATHA-like protein) were reported to be involved in seed size regulation. Mutant lines (*dpa4* and *sod7*) showed increased seed size. Similarly, ABA-induced transcription repressors (AITRs) were reported to be involved in the regulation of ABA signalling and abiotic stress tolerance. Arabidopsis *aitr2aitr5aitr6* (*aitr256*) triple mutant showed enhanced tolerance to drought and salt. Based on this information, quintuple mutants were generated using CRISPR/Cas9 approach and the mutants exhibited increased seed size and enhanced drought tolerance. This study also exhibited the role of DPA4 and SOD7 repressors in inflorescence architecture in Arabidopsis. A list of genes characterised by abiotic stress tolerance using genome engineering approaches is listed in Table 5.1. Various

Table 5.3 List of genes characterised using CRISPR genome editing

Crop	Gene/QTLs	Symbol (Gene/QTL)	Method	Target trait	References
Arabidopsis	*UDP-glycosyltransferases*	*UGT79B2 and UGT79B3*	CRISPR/Cas9	Abiotic stress	Li et al. (2017)
Arabidopsis	*C-repeat-binding factor*	*CBF1, CBF2, and CBF3*	CRISPR/Cas9	Cold	Cho et al. (2017)
Arabidopsis	*UDP-glycosyltransferases*	*UGT79B2, UGT79B3*	CRISPR/Cas9	Cold, salt, drought	Zhao and Zhu (2016)
Arabidopsis	*microRNA169a*	*MIR169a,*	CRISPR/Cas9	Drought	Zhao et al. (2016)
Arabidopsis	*OPEN STOMATA 2*	*OST2*	CRISPR/Cas9	Drought	Osakabe et al. (2016)
Rice	*Phytoene desaturase gene, Mitogen-activated protein kinase, betaine aldehyde dehydrogenase*	*OSPDS, OsMPK2, OsBADH2*	CRISPR/Cas9	Abiotic stress tolerance	Shan et al. (2013)
Rice	*Mitogen-activated protein kinase*	*OsMPK5*	CRISPR/Cas9	Abiotic stress tolerance	Xie and Yang (2013)
Rice	*Oryza sativa alternative oxidase 1 (AOX1)*	*OsAOX1a, OsAOX1b, OsAOX1c, OsBEL*	CRISPR/Cas9	Abiotic stress tolerance	Xu et al. (2015)
Rice	*calcium-dependent lipid binding annexin*	*OsAnn3*	CRISPR/Cas9	Cold	Shen et al. (2017)
Rice	*osmotic stress/ABA–activated protein kinase 2*	*OsSAPK2*	CRISPR/Cas9	Drought	Lou et al. (2017)
Rice	*Oryza sativa ethylene response factor (ERF) gene (OsDERF1); Oryza sativa photo-period sensitive male sterile (OsPMS3)*	*OsDERF1, OsPMS3, OsEPSPS, OsMSH1, OsMYB5*	CRISPR/Cas9	Drought	Zhang et al. (2014)
Rice	*NAM, ATAF and CUC (NAC) transcription factors*	*OsNAC041*	CRISPR/Cas9	Salinity	Bo et al. (2019)
Rice	*Mitogen-activated protein kinase*	*OsMPK2, OsDEP1*	CRISPR/Cas9	Yield under stress	Shan et al. (2014)
Potato	*Coilin gene*	*Coilin gene*	CRISPR-Cas9 RNP complex, CRISPR/Cas9	Abiotic stress	Khromov et al. (2018), Makhotenko et al. (2019)

(continued)

Table 5.3 (continued)

Crop	Gene/QTLs	Symbol (Gene/QTL)	Method	Target trait	References
Potato	*Acetolactate synthase*	*StALS1*, *StALS2*	CRISPR/Cas9 cytidine base editor	Herbicide	Veillet et al. (2019a)
Potato	*Acetolactate synthase*	*StALS1*	CRISPR/Cas9	Herbicide	Butler et al. (2015)
Potato	*Acetolactate synthase*	*StALS1*, *StALS2*	CRISPR/Cpf1	Herbicide	Veillet et al. (2019b)
Tomato	*C-repeat-binding factor*	*CBF1*	CRISPR/Cas9	Chilling tolerance	Li et al. (2018)
Tomato	*Stable tomatonon expressor of pathogenesis-related gene 1*	*SlNPR1*	CRISPR/Cas9	Drought	Li et al. (2019a, b)
Tomato	*Mitogen-activated protein kinases*	*SlMAPK3*	CRISPR/Cas9	Drought	Wang et al. (2017)
Tomato	*Acetolactate synthase*	*SlALS1*	CRISPR/Cas9	Herbicide	Danilo et al. (2019)
Tomato	*Acetolactate synthase*	*SlALS1*, *SlALS2*	CRISPR/Cpf1	Herbicide	Veillet et al. (2019a)
Watermelon	*Acetolactate synthase*	*ALS*	CRISPR/Cas9	Herbicide	Tian et al. (2018)
Wheat	*Dehydration responsive element binding*	*TaDREB2*, *TaERF3*	CRISPR/Cas9	Abiotic stress	Kim et al. (2018)

pathways and genes to be targeted for application of genome editing to improve abiotic stress tolerance in crops (Fig. 5.2).

5.4 Genome Engineering for Salinity Tolerance

Every single day, production loss of 2000 ha of arable land is reported globally due to salinisation (Shahid et al. 2018). Major causes of salinity are natural phenomenon such as seawater drift, or the result of water evaporation and transpiration causing the accumulation of salts in the soil. Accumulated salts interrupt the nutrients uptake of the root system and affect the plant growth. Human interventions such as luxurious usage of fertilisers, over-irrigation and irrigation of saline water also cause salinity. Plants exhibit three types of salinity tolerance mechanisms such as osmotic tolerance (signalling and sensing mechanism), ion tolerance (exclusion of ions via roots to avoid accumulation in leaves) and tissue tolerance (compartmentalisation of the accumulated salts in leaves at a cellular and intracellular levels) (Roy et al. 2014). Salinity tolerance is defined as the ability of a plant to balance biomass and/or yield

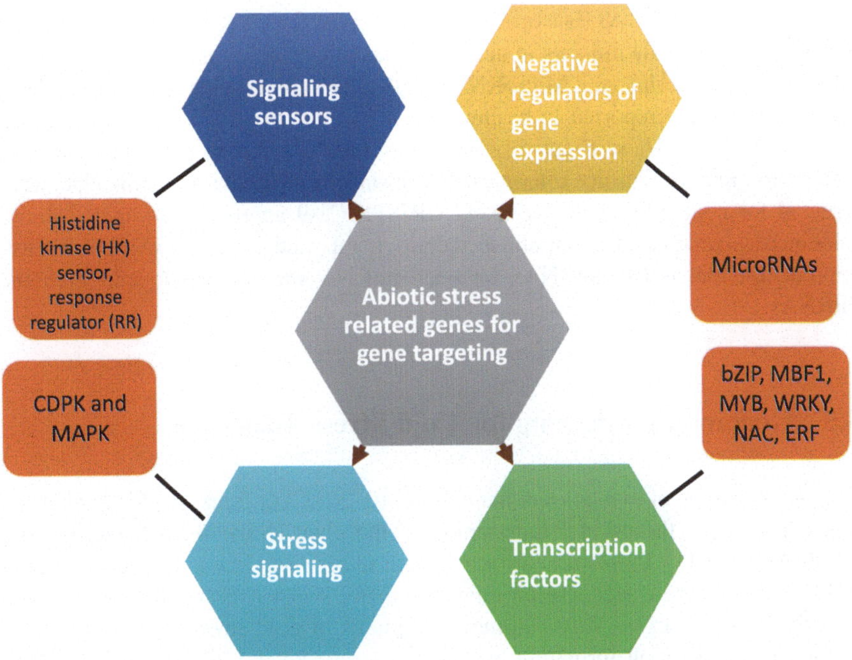

Fig. 5.2 Pathways and genes to be targeted for application of genome editing to improve abiotic stress tolerance in crops

under salt stress conditions (Schmöckel 2014). Several efforts through conventional breeding and transgenic approach were taken to improve rice productivity under high salinity.

OsRR22 in rice regulates cytokinin signal transduction and metabolism. Reduced level of *OsRR22* transcripts was reported to increase salinity tolerance (Takagi et al. 2015). A recent study by Zhang et al. (2019a) reported editing of *OsRR22* gene using CRISPR/Cas9 approach. Analysis of two homozygous T_2 lines showed enhanced salinity tolerance and no significant differences in other agronomic traits were observed between the edited lines and wild type lines. In order to study the regulatory mechanism of H_2O_2 signalling and K^+ uptake in root under salt stress (Huang et al. 2019) studied the transcriptome of salt-sensitive cucumber and salt-tolerant pumpkin. Higher expression of NADPH oxidase (respiratory burst oxidase homolog D; RBOHD), 14-3-3 protein (GRF12), plasma membrane H^+-ATPase (AHA1), and potassium transporter (HAK5) was observed in a pumpkin than in cucumber. CRISPR/Cas9 mediated knocking out of the coding sequences of NADPH oxidase (respiratory burst oxidase homolog D; RBOHD) showed reduced root apex H_2O_2 and K^+ content and GRF12, AHA1, and HAK5 expression, leading to a salt-sensitive phenotype. This study showed that RBOHD dependent H_2O_2 signalling in the root apex is important for pumpkin salt tolerance. Further, this study demonstrated the RBOHD-mediated transcriptional and post-translational activation of plasma membrane H+-ATPase operating upstream of HAK5 K^+ uptake transporters. Studying the RBOHD mutants in detail will pave the way in understanding salinity tolerance in *cucurbitaceae*. MicroRNAs (miRNA) play a major regulatory role in plants and few are reported to be involved in salinity stress tolerance mechanism (Gao et al. 2011). These negatively regulating miRNAs can be employed for gene editing to enhance salinity tolerance. For instance, miR393a and miR396c were targeted for gene editing in rice using CRISPR/Cas9 approach to impart salinity tolerance (Jaganathan et al. unpublished data). In this study, multiplex CRISPR/Cas9 approaches, that is two sgRNAs for each miRNA was applied to edit both the miRNAs.

5.5 Genome Engineering for Cold Stress Tolerance

Cold stress can be of two types either chilling (0–20 °C) or freezing (0 °C or below) which can affect the normal functioning of the plant (Kazemi-Shahandashti and Maali-Amiri 2018). Plants feel the stress both in high and low temperature (Yadav 2010). Crops such as rice, cotton, maize, tomato, sweet potato and cucumber are sensitive to cold stress. Very common symptoms of cold stress are rapid wilting followed by sunken pit formation leading to necrotic patches of tissues on the lead surface. When plants are exposed to low temperature, water stress is the major sign as leaf water decreases due to a decrease in root hydraulic conductance leading to rendered growth (Rasool et al. 2015).

Li et al. (2018) demonstrated the molecular basis of cold tolerance in tomato using CRISPR/Cas9 based knock out approach. C-repeat binding factors (CBFs) are reported to be involved in cold stress response in many species (Zhao et al. 2017). This study targeted to knock out the *slCBF* in tomato. The *slcbf* mutants showed severe chilling injuries than the wild type plants. Similarly, *OsAnn3* knockouts generated through CRISPR/Cas9 approach in rice showed reduced chilling tolerance (Shen et al. 2017). This study suggested the role of *OsAnn3* in cold tolerance of rice. Extensive work had been done in order to understand the mechanism of cold stress tolerance in crop plants. Such studies identified several important cold stress tolerances related genes including, transcription factor *viz OsWRKY11*, Hv-WRKY38, *ICE1*, *HSP70* through plant genetic transformation (Mare et al. 2004; Wu et al. 2009; Zuo et al. 2019; Zhao et al. 2019). These genes can be targeted for gene editing using CRISPR approach and validated for cold tolerance.

5.6 Crispr/Cas9 Genome Engineering for Heat Stress Tolerance

High temperature is fast becoming a major abiotic stress affecting plant growth, morphology and productivity (Hemmati et al. 2015). Heat stress is defined as hot temperatures beyond a threshold level for a period of time resulting in irreversible damage to plant growth and development (Hall 2012). Generally, temperature above 10–15 °C is considered heat shock or heat stress. Plants express a variety of responses to high temperatures. Though heat stress affects plant growth at all developmental stages, later phenological stages in particular anthesis and grain filling, are generally more susceptible (Hasanuzzaman et al. 2013). Temperature stress is an emerging concern with regard to global food security (Jha et al. 2015). Hence, addressing heat stress through biotechnological approaches is the need of the hour.

Genetic engineering for high-temperature tolerance in plants can be achieved either by over-expressing genes of heat shock protein or altering levels of heat shock transcription factor as well as by elevating levels of osmolytes. Over the period of research, it had been concluded that *Hsps* helps in minimising the damage to cell proteins hence improving the heat tolerance. Successful examples of enhancement of the high temperatures tolerance includes genetic engineering of glycine betaine synthesis in Arabidopsis (Hayashi et al. 1998), or suppression of *OsMDHAR4* in rice (Liu et al. 2018); expression of rice heat stress transcription factor *OsHsfA2e* in transgenic Arabidopsis (Yokotani et al. 2008). Similarly reports on the expression of transcription factor *OsbZIP46CA1* and SAPK6 (protein kinase) resulted in improved tolerance to heat and cold stress in rice (Chang et al. 2017). Recent reports on rice indicated that overexpression of protein disulfide isomerase gene from methanothermobacter thermautitrophicus enhanced heat tolerance in rice. Furthermore, Yu et al. (2017) have demonstrated that overexpression of poplar genes *PtPYRL1* and *PtPYRL5* enhances tolerance to drought and cold stresses by activating the ABA signalling pathway. Esmaeili et al. (2019) had proven enhanced tolerance to drought, salt, and heat stresses byco-overexpression of *AVP1/OsSIZ1* and provides the proof-of-concept that double gene overexpression performs significantly better than *AVP1*-overexpressing plants and *OsSIZ1*-overexpressing plants.

Manipulation of genes, transcription factors and signalling/metabolic pathways associated with cold and heat tolerance can effectively generate-tolerant crop varieties. Genome editing can be a valuable option for rapid and effective generation of abiotic stress-tolerant crop varieties. Few successful reports using CRISPR/Cas9 genome editing to confer abiotic stress tolerance were demonstrated in Arabidopsis, Rice and other crops. Miao et al. (2018) had worked on the *OsPYL* (abscisic acid receptor gene family) and successfully created *pyl1/4/6* triple knockout rice by CRISPR/Cas9 editing, resulted in increased grain yield, greater high-temperature tolerance.

5.7 Major Challenges for the Application of Genome Editing in Crops

(i) *Off targets*
Though several attempts have been made to reduce the off-target effects in the gene-edited plants such as identification of cas9 orthologous and cas9 variants, still, the specificity issues persist. It comes to more concern while considering medical applications than in crop plants.
(ii) *Protocol optimisation*
It is difficult to develop a suitable tissue culture protocol for crops such as cotton, tree species including coconut, citrus, cashew, pomegranate, mango and woody species etc. Several studies are being conducted to improve tissue culture protocol for these species and applied to genome editing (Jia et al. 2019).
(iii) *Public acceptance*
Although genome editing has obtained more popularity among the science community and applied to many crop systems for developing improved lines, all the research is at the laboratory level. Regulatory measures to determine the gene-edited crops as transgene-free crops or genetically modified crops is still uncertain in many countries and there is no internationally accepted frameworks are available in the current scenario (Mao et al. 2019).

5.8 Conclusion

Development of agrotechniques is the need of the hour to cope up the food production with the ever-growing population, extreme weather changes and temperature increase. Though improvements in conventional breeding and tissue culture technologies exist, these approaches were reported to have their own limitations; inefficient in genetic manipulation of complex traits, time-consuming and linkage drag. The application of genome editing tool (CRISPR/Cas9) has been applied for enhancing abiotic stress tolerance such as drought in tomato, cold tolerance and high temperature in rice. CRISPR/cas9 can be improved by optimisation of the promoter or by utilisation of reporter or selectable markers. The application of genome editing had proven its significance in genome editing and crop improvement. Although CRISPR/Cas can be potential technology, precise customised genome editing will be needed to significantly boost crop research and agricultural industries.

References

Abebe T, Guenzi AC, Martin B, Cushman JC (2003) Tolerance of mannitol-accumulating transgenic wheat to water stress and salinity. Plant Physiol 131:1748–1755

Allen RD (1995) Dissection of oxidative stress tolerance using transgenic plants. Plant Physiol 107:1049

Almeida GD, Nair S, Borém A, Cairns J, Trachsel S, Ribaut J-M, Bänziger M, Prasanna BM, Crossa J, Babu R (2014) Molecular mapping across three populations reveals a QTL hotspot region on chromosome 3 for secondary traits associated with drought tolerance in tropical maize. Mol Breed 34:701–715

Babu V, Bansal K (1998) Osmotin overexpression in transgenic potato plants provide protection against osmotic stress. In: 5th international symposium on molecular biology of potato, Bogensee, Germany, 1998, pp 2–6

Babu RC, Nguyen BD, Chamarerk V, Shanmugasundaram P, Chezhian P, Jeyaprakash P, Ganesh S, Palchamy A, Sadasivam S, Sarkarung S (2003) Genetic analysis of drought resistance in rice by molecular markers. Crop Sci 43:1457–1469

Bortesi L, Fischer R (2015) The CRISPR/Cas9 system for plant genome editing and beyond. Biotechnol Adv 33:41–52

Bo W, Zhaohui Z, Huanhuan Z, Xia W, Binglin L, Lijia Y, Xiangyan H, Deshui Y, Xuelian Z, Chunguo W (2019) Targeted mutagenesis of NAC transcription factor gene, OsNAC041, leading to salt sensitivity in rice. Rice Sci 26:98–108

Butler NM, Atkins PA, Voytas DF, Douches DS (2015) Generation and inheritance of targeted mutations in potato (Solanum tuberosum L.) using the CRISPR/Cas system. PloS One 10:e0144591

Cai S, Jiang G, Ye N, Chu Z, Xu X, Zhang J, Zhu G (2015) A key ABA catabolic gene, OsABA8ox3, is involved in drought stress resistance in rice. PLoS ONE 10:e0116646

Chang Y, Nguyen BH, Xie Y, Xiao B, Tang N, Zhu W, Mou T, Xiong L (2017) Co-overexpression of the constitutively active form of OsbZIP46 and ABA-activated protein kinase SAPK6 improves drought and temperature stress resistance in rice. Front Plant Sci 8:1102

Cheng Z, Targolli J, Huang X, Wu R (2002) Wheat LEA genes, PMA80 and PMA1959, enhance dehydration tolerance of transgenic rice (Oryza sativa L.). Mol Breed 10:71–82

Chen W, Li W (2005) Mapping of QTL conferring cold tolerance at early seedling stage of rice by molecular markers. Wuhan Bot Res 23:116–120

Chen K, Wang Y, Zhang R, Zhang H, Gao C (2019) CRISPR/Cas genome editing and precision plant breeding in agriculture. Annu Rev Plant Biol 70:667–697

Chin JH, Lu X, Haefele SM, Gamuyao R, Ismail A, Wissuwa M, Heuer S (2010) Development and application of gene-based markers for the major rice QTL phosphorus uptake 1. Theor Appl Genet 120:1073–1086

Cho EK, Hong CB (2006) Over-expression of tobacco NtHSP70-1 contributes to drought-stress tolerance in plants. Plant Cell Rep 25:349–358

Cho S, Kumar J, Shultz JL, Anupama K, Tefera F, Muehlbauer FJ (2002) Mapping genes for double podding and other morphological traits in chickpea. Euphytica 128:285–292

Cho JS, Choi KR, Prabowo CPS, Shin JH, Yang D, Jang J, Lee SY (2017) CRISPR/Cas9-coupled recombineering for metabolic engineering of Corynebacterium glutamicum. Metab Eng 42:157–167

Chukwu S, Rafii M, Ramlee S, Ismail S, Hasan M, Oladosu Y, Magaji U, Akos I, Olalekan K (2019) Bacterial leaf blight resistance in rice: a review of conventional breeding to molecular approach. Mol Biol Rep 46:1519–1532

Danilo B, Perrot L, Mara K, Botton E, Nogué F, Mazier M (2019) Efficient and transgene-free gene targeting using Agrobacterium-mediated delivery of the CRISPR/Cas9 system in tomato. Plant Cell Rep 38:459–462

Daspute AA, Kobayashi Y, Panda SK, Fakrudin B, Kobayashi Y, Tokizawa M, Iuchi S, Choudhary AK, Yamamoto YY, Koyama H (2018) Characterization of CcSTOP1; a C2H2-type transcription factor regulates Al tolerance gene in pigeonpea. Planta 247:201–214

Dhillon T, Pearce SP, Stockinger EJ, Distelfeld A, Li C, Knox AK, Vashegyi I, Vágújfalvi A, Galiba G, Dubcovsky J (2010) Regulation of freezing tolerance and flowering in temperate cereals: the VRN-1 connection. Plant Physiol 153:1846–1858

Emon RM, Islam MM, Halder J, Fan Y (2015) Genetic diversity and association mapping for salinity tolerance in Bangladeshi rice landraces. Crop J 3:440–444

Esmaeili N, Yang X, Cai Y, Sun L, Zhu X, Shen G, Payton P, Fang W, Zhang H (2019) Co-overexpression of AVP1 and OsSIZ1 in Arabidopsis substantially enhances plant tolerance to drought, salt, and heat stresses. Sci Rep 9:7642

Francia E, Rizza F, Cattivelli L, Stanca AM, Galiba G, Toth B, Hayes PM, Skinner JS, Pecchioni N (2004) Two loci on chromosome 5H determine low-temperature tolerance in a 'Nure'(winter) × 'Tremois' (spring) barley map. Theor Appl Genet 108:670–680

Francia E, Barabaschi D, Tondelli A, Laidò G, Rizza F, Stanca AM, Busconi M, Fogher C, Stockinger EJ, Pecchioni N (2007) Fine mapping of a HvCBF gene cluster at the frost resistance locus Fr-H2 in barley. Theor Appl Genet 115:1083–1091

Gahlaut V, Jaiswal V, Tyagi BS, Singh G, Sareen S, Balyan HS, Gupta PK (2017) QTL mapping for nine drought-responsive agronomic traits in bread wheat under irrigated and rain-fed environments. PLoS ONE 12:e0182857

Gamuyao R, Chin JH, Pariasca-Tanaka J, Pesaresi P, Catausan S, Dalid C, Slamet-Loedin I, Tecson-Mendoza EM, Wissuwa M, Heuer S (2012) The protein kinase Pstol1 from traditional rice confers tolerance of phosphorus deficiency. Nature 488:535

Gao S, Yuan L, Zhai H, Liu C, He S, Liu Q (2011) Transgenic sweetpotato plants expressing an LOS5 gene are tolerant to salt stress. Plant Cell Tissue Organ Cult (PCTOC) 107:205

Garg AK, Kim J-K, Owens TG, Ranwala AP, DO CHOI, Y., KOCHIAN, L. V. & WU, R. J. (2002) Trehalose accumulation in rice plants confers high tolerance levels to different abiotic stresses. Proc Natl Acad Sci 99:15898–15903

Gaudelli NM, Komor AC, Rees HA, Packer MS, Badran AH, Bryson DI, Liu DR (2017) Programmable base editing of A• T to G• C in genomic DNA without DNA cleavage. Nature 551:464

Hall AE (2012) 5 heat stress. Plant Stress Physiol 118

Hao Z, Li X, Liu X, Xie C, Li M, Zhang D, Zhang S (2010) Meta-analysis of constitutive and adaptive QTL for drought tolerance in maize. Euphytica 174:165–177

Hasanuzzaman M, Nahar K, Alam M, Roychowdhury R, Fujita M (2013) Physiological, biochemical, and molecular mechanisms of heat stress tolerance in plants. Int J Mol Sci 14:9643–9684

Hattori Y, Nagai K, Furukawa S, Song X-J, Kawano R, Sakakibara H, Wu J, Matsumoto T, Yoshimura A, Kitano H (2009) The ethylene response factors SNORKEL1 and SNORKEL2 allow rice to adapt to deep water. Nature 460:1026

Hayashi H, Sakamoto A, Murata N (1998) Enhancement of the tolerance of Arabidopsis to high temperatures by genetic engineering of the synthesis of glycinebetaine. Plant J 16:155–161

Hemmati H, Gupta D, Basu C (2015) Molecular physiology of heat stress responses in plants. Elucidation of abiotic stress signaling in plants. Springer.

Hoshida H, Tanaka Y, Hibino T, Hayashi Y, Tanaka A, Takabe T, Takabe T (2000) Enhanced tolerance to salt stress in transgenic rice that overexpresses chloroplast glutamine synthetase. Plant Mol Biol 43:103–111

Hsieh T-H, Lee J-T, Charng Y-Y, Chan M-T (2002) Tomato plants ectopically expressing Arabidopsis CBF1 show enhanced resistance to water deficit stress. Plant Physiol 130:618–626

Huang Y, Cao H, Yang L, Chen C, Shabala L, Xiong M, Niu M, Liu J, Zheng Z, Zhou L (2019) Tissue-specific respiratory burst oxidase homolog-dependent H2O2 signaling to the plasma membrane H+-ATPase confers potassium uptake and salinity tolerance in Cucurbitaceae. J Exp Bot 70:5879–5893

Hu H, Dai M, Yao J, Xiao B, Li X, Zhang Q, Xiong L (2006) Overexpressing a NAM, ATAF, and CUC (NAC) transcription factor enhances drought resistance and salt tolerance in rice. Proc Natl Acad Sci 103:12987–12992

Jaganathan D, Thudi M, Kale S, Azam S, Roorkiwal M, Gaur PM, Kishor PK, Nguyen H, Sutton T, Varshney RK (2015) Genotyping-by-sequencing based intra-specific genetic map refines a "QTL-hotspot" region for drought tolerance in chickpea. Mol Genet Genomics 290:559–571

Jaganathan D, Ramasamy K, Sellamuthu G, Jayabalan S, Venkataraman G (2018) CRISPR for crop improvement: an update review. Front Plant Sci 9

Jaganathan D, Abhishek B, Mahendar T, Varshney R (2020) Fine mapping and gene cloning in the post-NGS era: advances and prospects. Theor Appl Genet

James RA, Blake C, Zwart AB, Hare RA, Rathjen AJ, Munns R (2012) Impact of ancestral wheat sodium exclusion genes Nax1 and Nax2 on grain yield of durum wheat on saline soils. Funct Plant Biol 39:609–618

Jang I-C, Oh S-J, Seo J-S, Choi W-B, Song SI, Kim CH, Kim YS, Seo H-S, Do Choi Y, Nahm BH (2003) Expression of a bifunctional fusion of the Escherichia coli genes for trehalose-6-phosphate synthase and trehalose-6-phosphate phosphatase in transgenic rice plants increases trehalose accumulation and abiotic stress tolerance without stunting growth. Plant Physiol 131:516–524

Jena K, Kim S, Suh J, Ki Y (2010) Development of cold-tolerant breeding lines using QTL analysis in rice. In: Second Africa rice congress, Bamako, 2010, pp 22–26

Jeong M-J, Park S-C, Byun M-O (2001) Improvement of salt tolerance in transgenic potato plants by glyceraldehyde-3 phosphate dehydrogenase gene transfer. Mol Cells 12. Springer Science & Business Media BV

Jha A, Jeong D-W, Jang W-J, Rode CV, Roh H-S (2015) Mesoporous NiCu–CeO 2 oxide catalysts for high-temperature water–gas shift reaction. RSC Adv 5:1430–1437

Jiang W, Zhou H, Bi H, Fromm M, Yang B, Weeks DP (2013) Demonstration of CRISPR/Cas9/sgRNA-mediated targeted gene modification in Arabidopsis, tobacco, sorghum and rice. Nucleic Acids Res 41:e188–e188

Jia H, Zou X, Orbovic V, Wang N (2019) Genome editing in citrus tree with CRISPR/Cas9. Plant genome editing with CRISPR systems. Springer

Jinek M, Chylinski K, Fonfara I, Hauer M, Doudna JA, Charpentier E (2012) A programmable dual-RNA–guided DNA endonuclease in adaptive bacterial immunity. Science 337:816–821

Kamoshita A, Babu RC, Boopathi NM, Fukai S (2008) Phenotypic and genotypic analysis of drought-resistance traits for development of rice cultivars adapted to rainfed environments. Field Crops Res 109:1–23

Kazemi-Shahandashti S-S, Maali-Amiri R (2018) Global insights of protein responses to cold stress in plants: signaling, defence, and degradation. J Plant Physiol 226:123–135

Khowaja FS, Norton GJ, Courtois B, Price AH (2009) Improved resolution in the position of drought-related QTLs in a single mapping population of rice by meta-analysis. BMC Genomics 10:276

Khromov A, Gushchin V, Timerbaev V, Kalinina N, Taliansky M, Makarov V (2018) Guide RNA design for CRISPR/Cas9-mediated potato genome editing. Dokl Biochem Biophys 90–94. Springer

Kidokoro S, Watanabe K, Ohori T, Moriwaki T, Maruyama K, Mizoi J, Myint Phyu Sin Htwe N, Fujita Y, Sekita S, Shinozaki K (2015) Soybean DREB 1/CBF-type transcription factors function in heat and drought as well as cold stress-responsive gene expression. Plant J 81:505–518

Kim H, Lee K, Hwang H, Bhatnagar N, Kim D-Y, Yoon IS, Byun M-O, Kim ST, Jung K-H, Kim B-G (2014) Overexpression of PYL5 in rice enhances drought tolerance, inhibits growth, and modulates gene expression. J Exp Bot 65:453–464

Kim D, Alptekin B, Budak H (2018) CRISPR/Cas9 genome editing in wheat. Funct Integr Genomics 18:31–41

Kirigwi F, van Ginkel M, Brown-Guedira G, Gill B, Paulsen GM, Fritz A (2007) Markers associated with a QTL for grain yield in wheat under drought. Mol Breed 20:401–413

Knox AK, Dhillon T, Cheng H, Tondelli A, Pecchioni N, Stockinger EJ (2010) CBF gene copy number variation at frost resistance-2 is associated with levels of freezing tolerance in temperate-climate cereals. Theor Appl Genet 121:21–35

Komor AC, Kim YB, Packer MS, Zuris JA, Liu DR (2016) Programmable editing of a target base in genomic DNA without double-stranded DNA cleavage. Nature 533:420

Kumar A, Dixit S, Ram T, Yadaw R, Mishra K, Mandal N (2014) Breeding high-yielding drought-tolerant rice: genetic variations and conventional and molecular approaches. J Exp Bot 65:6265–6278

Lafitte H, Yan S, Gao Y, Li Z-K (2004) Response to direct selection for yield under reproductive stage stress in rice backcross populations. Resilient crops for water limited environments, vol 169

Laurie S, Feeney KA, Maathuis FJ, Heard PJ, Brown SJ, Leigh RA (2002) A role for HKT1 in sodium uptake by wheat roots. Plant J 32:139–149

Lee CM, Cradick TJ, Bao G (2016) The Neisseria meningitidis CRISPR-Cas9 system enables specific genome editing in mammalian cells. Mol Ther 24:645–654

Lichtenzveig J, Bonfil DJ, Zhang H-B, Shtienberg D, Abbo S (2006) Mapping quantitative trait loci in chickpea associated with time to flowering and resistance to Didymella rabiei the causal agent of Ascochyta blight. Theor Appl Genet 113:1357–1369

Liu X, Hong L, Li X-Y, Yao Y, Hu B, Li L (2011) Improved drought and salt tolerance in transgenic Arabidopsis overexpressing a NAC transcriptional factor from Arachis hypogaea. Biosci Biotechnol Biochem 75:443–450

Liu J, Sun X, Xu F, Zhang Y, Zhang Q, Miao R, Zhang J, Liang J, Xu W (2018) Suppression of OsMDHAR4 enhances heat tolerance by mediating H 2 O 2-induced stomatal closure in rice plants. Rice 11:38

Li J-Y, Liu J, Dong D, Jia X, McCouch SR, Kochian LV (2014) Natural variation underlies alterations in Nramp aluminum transporter (NRAT1) expression and function that play a key role in rice aluminum tolerance. Proc Natl Acad Sci 111:6503–6508

Li C, Sun B, Li Y, Liu C, Wu X, Zhang D, Shi Y, Song Y, Buckler ES, Zhang Z (2016) Numerous genetic loci identified for drought tolerance in the maize nested association mapping populations. BMC Genomics 17:894

Li P, Li YJ, Zhang FJ, Zhang GZ, Jiang XY, Yu HM, Hou BK (2017) The Arabidopsis UDP-glycosyltransferases UGT79B2 and UGT79B3, contribute to cold, salt and drought stress tolerance via modulating anthocyanin accumulation. Plant J 89:85–103

Li R, Zhang L, Wang L, Chen L, Zhao R, Sheng J, Shen L (2018) Reduction of tomato-plant chilling tolerance by CRISPR–Cas9-mediated SlCBF1 mutagenesis. J Agric Food Chem 66:9042–9051

Li B, Rui H, Li Y, Wang Q, Alariqi M, Qin L, Sun L, Ding X, Wang F, Zou J (2019a) Robust CRISPR/Cpf1 (Cas12a) mediated genome editing in allotetraploid cotton (G. hirsutum). Plant Biotechnol J

Li R. Liu C, Zhao R, Wang L, Chen L, Yu W, Zhang S, Sheng J, Shen L (2019b) CRISPR/Cas9-mediated *SlNPR1* mutagenesisreduces tomato plant drought tolerance. BMC Plant Biol 19:38.F

Lou D, Wang H, Liang G, Yu D (2017) OsSAPK2 confers abscisic acid sensitivity and tolerance to drought stress in rice. Front Plant Sci 8:993

Magalhaes JV, Liu J, Guimaraes CT, Lana UG, Alves VM, Wang Y-H, Schaffert RE, Hoekenga OA, Pineros MA, Shaff JE (2007) A gene in the multidrug and toxic compound extrusion (MATE) family confers aluminum tolerance in sorghum. Nat Genet 39:1156

Mahfouz MM (2017) Genome editing: the efficient tool CRISPR–Cpf1. Nature Plants 3:1–2

Makhotenko A, Khromov A, Snigir E, Makarova S, Makarov V, Suprunova T, Kalinina N, Taliansky M (2019) Functional analysis of coilin in virus resistance and stress tolerance of potato solanum tuberosum using CRISPR-Cas9 editing. Dokl Biochem Biophys 88–91. Springer

Mao Y, Botella JR, Liu Y, Zhu J-K (2019) Gene editing in plants: progress and challenges. National Science Review 6:421–437

Maqbool S, Zhong H, El-Maghraby Y, Ahmad A, Chai B, Wang W, Sabzikar R, Sticklen M (2002) Competence of oat (Avena sativa L.) shoot apical meristems for integrative transformation, inherited expression, and osmotic tolerance of transgenic lines containing hva1. Theor Appl Genet 105:201–208

Mare C, Mazzucotelli E, Crosatti C, Francia E, Cattivelli L (2004) Hv-WRKY38: a new transcription factor involved in cold-and drought-response in barley. Plant Mol Biol 55:399–416

Maron LG, Guimarães CT, Kirst M, Albert PS, Birchler JA, Bradbury PJ, Buckler ES, Coluccio AE, Danilova TV, Kudrna D (2013) Aluminum tolerance in maize is associated with higher MATE1 gene copy number. Proc Natl Acad Sci 110:5241–5246

Mathews KL, Malosetti M, Chapman S, McIntyre L, Reynolds M, Shorter R, van Eeuwijk F (2008) Multi-environment QTL mixed models for drought stress adaptation in wheat. Theor Appl Genet 117:1077–1091

Matsumura T, Tabayashi N, Kamagata Y, Souma C, Saruyama H (2002) Wheat catalase expressed in transgenic rice can improve tolerance against low temperature stress. Physiol Plant 116:317–327

McKersie BD, Chen Y, de Beus M, Bowley SR, Bowler C, Inzé D, D'Halluin K, Botterman J (1993) Superoxide dismutase enhances tolerance of freezing stress in transgenic alfalfa (Medicago sativa L.). Plant Physiol 103:1155–1163

McKersie BD, Bowley SR, Harjanto E, Leprince O (1996) Water-deficit tolerance and field performance of transgenic alfalfa overexpressing superoxide dismutase. Plant Physiol 111:1177–1181

McKersie BD, Bowley SR, Jones KS (1999) Winter survival of transgenic alfalfa overexpressing superoxide dismutase. Plant Physiol 119:839–848

Miao C, Xiao L, Hua K, Zou C, Zhao Y, Bressan RA, Zhu J-K (2018) Mutations in a subfamily of abscisic acid receptor genes promote rice growth and productivity. Proc Natl Acad Sci 115:6058–6063

Mohanty A, Kathuria H, Ferjani A, Sakamoto A, Mohanty P, Murata N, Tyagi A (2002) Transgenics of an elite indica rice variety Pusa Basmati 1 harbouring the codA gene are highly tolerant to salt stress. Theor Appl Genet 106:51–57

Munns R, James RA, Xu B, Athman A, Conn SJ, Jordans C, Byrt CS, Hare RA, Tyerman SD, Tester M (2012) Wheat grain yield on saline soils is improved by an ancestral Na+ transporter gene. Nat Biotechnol 30:360

Neeraja C, Mishra B, Rao KS, Singh R, Padmavati G, Shenoy V (2008) Linkage disequilibrium in salt tolerant genotypes of rice (Oryza sativa L). J Plant Biochem Biotechnol 17:65–68

Osakabe Y, Watanabe T, Sugano SS, Ueta R, Ishihara R, Shinozaki K, Osakabe K (2016) Optimization of CRISPR/Cas9 genome editing to modify abiotic stress responses in plants. Sci Rep 6:26685

Paixão JFR, Gillet F-X, Ribeiro TP, Bournaud C, Lourenço-Tessutti IT, Noriega DD, de Melo BP, De Almeida-Engler J, Grossi-De-sa MF (2019) Improved drought stress tolerance in Arabidopsis by CRISPR/dCas9 fusion with a Histone AcetylTransferase. Sci Rep 9:8080

Payton P, Webb R, Kornyeyev D, Allen R, Holaday AS (2001) Protecting cotton photosynthesis during moderate chilling at high light intensity by increasing chloroplastic antioxidant enzyme activity. J Exp Bot 52:2345–2354

Pazhamala L, Saxena RK, Singh VK, Sameerkumar C, Kumar V, Sinha P, Patel K, Obala J, Kaoneka SR, Tongoona P (2015) Genomics-assisted breeding for boosting crop improvement in pigeonpea (Cajanus cajan). Front Plant Sci 6:50

Perl A, Perl-Treves R, Galili S, Aviv D, Shalgi E, Malkin S, Galun E (1993) Enhanced oxidative-stress defense in transgenic potato expressing tomato Cu, Zn superoxide dismutases. Theor Appl Genet 85:568–576

Pilon-Smits EA, Zhu YL, Sears T, Terry N (2000) Overexpression of glutathione reductase in Brassica juncea: effects on cadmium accumulation and tolerance. Physiol Plant 110:455–460

Pinto RS, Reynolds MP, Mathews KL, McIntyre CL, Olivares-Villegas J-J, Chapman SC (2010) Heat and drought adaptive QTL in a wheat population designed to minimize confounding agronomic effects. Theor Appl Genet 121:1001–1021

Priyanka B, Sekhar K, Sunita T, Reddy V, Rao KV (2010) Characterization of expressed sequence tags (ESTs) of pigeonpea (Cajanus cajan L.) and functional validation of selected genes for abiotic stress tolerance in Arabidopsis thaliana. Mol Genet Genomics 283:273–287

Rahman H, Dakshinamurthi V, Ramasamy S, Manickam S, Kaliyaperumal AK, Raha S, Panneerselvam N, Ramanathan V, Nallathambi J, Sabariappan R (2018) Introgression of submergence tolerance into CO 43, a popular rice variety of India, through marker-assisted backcross breeding. Czech J Genet Plant Breed 54:101–108

Ran FA, Cong L, Yan WX, Scott DA, Gootenberg JS, Kriz AJ, Zetsche B, Shalem O, Wu X, Makarova KS (2015) In vivo genome editing using Staphylococcus aureus Cas9. Nature 520:186

Rasool S, Abdel Latef A, Ahmad P (2015) Chickpea: role and responses under abiotic and biotic stress. Legumes under environmental stress: yield, improvement and adaptations, pp 67–79

Ray DK, Mueller ND, West PC, Foley JA (2013) Yield trends are insufficient to double global crop production by 2050. PLoS ONE 8:e66428

Rehman A, Malhotra R, Bett K, Tar'An B, Bueckert R, Warkentin T (2011) Mapping QTL associated with traits affecting grain yield in chickpea (Cicer arietinum L.) under terminal drought stress. Crop Sci 51:450–463

Ren Z-H, Gao J-P, Li L-G, Cai X-L, Huang W, Chao D-Y, Zhu M-Z, Wang Z-Y, Luan S, Lin H-X (2005) A rice quantitative trait locus for salt tolerance encodes a sodium transporter. Nat Genet 37:1141

Rohila JS, Jain RK, Wu R (2002) Genetic improvement of Basmati rice for salt and drought tolerance by regulated expression of a barley Hva1 cDNA. Plant Sci 163:525–532

Roy M, Wu R (2001) Arginine decarboxylase transgene expression and analysis of environmental stress tolerance in transgenic rice. Plant Sci 160:869–875

Roy M, Wu R (2002) Overexpression of S-adenosylmethionine decarboxylase gene in rice increases polyamine level and enhances sodium chloride-stress tolerance. Plant Sci 163:987–992

Roy SJ, Negrão S, Tester M (2014) Salt resistant crop plants. Curr Opin Biotechnol 26:115–124

Saijo Y, Hata S, Kyozuka J, Shimamoto K, Izui K (2000) Over-expression of a single Ca2+-dependent protein kinase confers both cold and salt/drought tolerance on rice plants. Plant J 23:319–327

Sakuma Y, Maruyama K, Qin F, Osakabe Y, Shinozaki K, Yamaguchi-Shinozaki K (2006) Dual function of an Arabidopsis transcription factor DREB2A in water-stress-responsive and heat-stress-responsive gene expression. Proc Natl Acad Sci 103:18822–18827

Sasaki T, Yamamoto Y, Ezaki B, Katsuhara M, Ahn SJ, Ryan PR, Delhaize E, Matsumoto H (2004) A wheat gene encoding an aluminum-activated malate transporter. Plant J 37:645–653

Sasaki T, Ryan PR, Delhaize E, Hebb DM, Ogihara Y, Kawaura K, Noda K, Kojima T, Toyoda A, Matsumoto H (2006) Sequence upstream of the wheat (Triticum aestivum L.) ALMT1 gene and its relationship to aluminum resistance. Plant Cell Physiol 47:1343–1354

Schmöckel SM (2014) Salinity detection and control of sodium transport in Arabidopsis thaliana

Schwartz SH, Qin X, Zeevaart JA (2003) Elucidation of the indirect pathway of abscisic acid biosynthesis by mutants, genes, and enzymes. Plant Physiol 131:1591–1601

Shahid SA, Zaman M, Heng L (2018) Introduction to soil salinity, sodicity and diagnostics techniques. Guideline for salinity assessment, mitigation and adaptation using nuclear and related techniques. Springer

Shan Q, Wang Y, Li J, Zhang Y, Chen K, Liang Z, Zhang K, Liu J, Xi JJ, Qiu J-L (2013) Targeted genome modification of crop plants using a CRISPR-Cas system. Nat Biotechnol 31:686

Shan Q, Wang Y, Li J, Gao C (2014) Genome editing in rice and wheat using the CRISPR/Cas system. Nat Protoc 9:2395

Shen C, Que Z, Xia Y, Tang N, Li D, He R, Cao M (2017) Knock out of the annexin gene OsAnn3 via CRISPR/Cas9-mediated genome editing decreased cold tolerance in rice. J Plant Biol 60:539–547

Sivamani E, Bahieldin A, Wraith JM, Al-Niemi T, Dyer WE, Ho T-HD, Qu R (2000) Improved biomass productivity and water use efficiency under water deficit conditions in transgenic wheat constitutively expressing the barley HVA1 gene. Plant Sci 155:1–9

Sivasakthi K, Thudi M, Tharanya M, Kale SM, Kholová J, Halime MH, Jaganathan D, Baddam R, Thirunalasundari T, Gaur PM (2018) Plant vigour QTLs co-map with an earlier reported QTL hotspot for drought tolerance while water saving QTLs map in other regions of the chickpea genome. BMC Plant Biol 18:29

Skinner JS, Szűcs P, von Zitzewitz J, Marquez-Cedillo L, Filichkin T, Stockinger EJ, Thomashow MF, Chen TH, Hayes PM (2006) Mapping of barley homologs to genes that regulate low temperature tolerance in Arabidopsis. Theor Appl Genet 112:832–842

Song G, Jia M, Chen K, Kong X, Khattak B, Xie C, Li A, Mao L (2016) CRISPR/Cas9: a powerful tool for crop genome editing. Crop J 4:75–82

Suji K, Biji K, Poornima R, Prince KSJ, Amudha K, Kavitha S, Mankar S, Babu RC (2012) Mapping QTLs for plant phenology and production traits using indica rice (Oryza sativa L.) lines adapted to rainfed environment. Mol Biotechnol 52:151–160

Swamy BM, Vikram P, Dixit S, Ahmed H, Kumar A (2011) Meta-analysis of grain yield QTL identified during agricultural drought in grasses showed consensus. BMC Genomics 12:319

Takagi H, Tamiru M, Abe A, Yoshida K, Uemura A, Yaegashi H, Obara T, Oikawa K, Utsushi H, Kanzaki E (2015) MutMap accelerates breeding of a salt-tolerant rice cultivar. Nat Biotechnol 33:445

Tang X, Lowder LG, Zhang T, Malzahn AA, Zheng X, Voytas DF, Zhong Z, Chen Y, Ren Q, Li Q (2017) A CRISPR–Cpf1 system for efficient genome editing and transcriptional repression in plants. Nat Plants 3:17018

Tayal D, Srivastava P, Bansal K (2003) Development of transgenic Indian mustard for abiotic stress tolerance. In: 2nd international congress of plant physiology on sustainable plant productivity under changing environment, p 479

Thomson MJ, de Ocampo M, Egdane J, Rahman MA, Sajise AG, Adorada DL, Tumimbang-Raiz E, Blumwald E, Seraj ZI, Singh RK (2010) Characterizing the Saltol quantitative trait locus for salinity tolerance in rice. Rice 3:148

Thornton PK (2012) Impacts of climate change on the agricultural and aquatic systems and natural resources within the CGIAR's mandate

Thudi M, Gaur PM, Krishnamurthy L, Mir RR, Kudapa H, Fikre A, Kimurto P, Tripathi S, Soren KR, Mulwa R (2014) Genomics-assisted breeding for drought tolerance in chickpea. Funct Plant Biol 41:1178–1190

Tian S, Jiang L, Cui X, Zhang J, Guo S, Li M, Zhang H, Ren Y, Gong G, Zong M (2018) Engineering herbicide-resistant watermelon variety through CRISPR/Cas9-mediated base-editing. Plant Cell Rep 37:1353–1356

Tilman D, Balzer C, Hill J, Befort BL (2011) Global food demand and the sustainable intensification of agriculture. Proc Natl Acad Sci 108:20260–20264

Trachsel S, Sun D, Sanvicente FM, Zheng H, Atlin GN, Suarez EA, Babu R, Zhang X (2016) Identification of QTL for early vigor and stay-green conferring tolerance to drought in two connected advanced backcross populations in tropical maize (Zea mays L.). PLoS One 11:e0149636

Uauy C, Distelfeld A, Fahima T, Blechl A, Dubcovsky J (2006) A NAC gene regulating senescence improves grain protein, zinc, and iron content in wheat. Science 314:1298–1301

Uga Y, Okuno K, Yano M (2011) Dro1, a major QTL involved in deep rooting of rice under upland field conditions. J Exp Bot 62:2485–2494

Uga Y, Sugimoto K, Ogawa S, Rane J, Ishitani M, Hara N, Kitomi Y, Inukai Y, Ono K, Kanno N (2013) Control of root system architecture by DEEPER ROOTING 1 increases rice yield under drought conditions. Nat Genet 45:1097

Valarmathi M, Sasikala R, Rahman H, Jagadeeshselvam N, Kambale R, Raveendran M (2019) Development of salinity tolerant version of a popular rice variety improved white ponni through marker assisted back cross breeding. Plant Physiology Reports 24:262–271

Varshney RK, Graner A, Sorrells ME (2005) Genic microsatellite markers in plants: features and applications. Trends Biotechnol 23:48–55

Varshney RK, Thudi M, Nayak SN, Gaur PM, Kashiwagi J, Krishnamurthy L, Jaganathan D, Koppolu J, Bohra A, Tripathi S (2014) Genetic dissection of drought tolerance in chickpea (Cicer arietinum L.). Theor Appl Genet 127:445–462

Varshney RK, Pandey MK, Bohra A, Singh VK, Thudi M, Saxena RK (2019) Toward the sequence-based breeding in legumes in the post-genome sequencing era. Theor Appl Genet 132:797–816

Veillet F, Perrot L, Chauvin L, Kermarrec M-P, Guyon-Debast A, Chauvin J-E, Nogué F, Mazier M (2019b) Transgene-free genome editing in tomato and potato plants using agrobacterium-mediated delivery of a CRISPR/Cas9 cytidine base editor. Int J Mol Sci 20:402

Veillet F, Chauvin L, Kermarrec M-P, Sevestre F, Merrer M, Terret Z, Szydlowski N, Devaux P, Gallois J-L, Chauvin J-E (2019a) The Solanum tuberosum GBSSI gene: a target for assessing gene and base editing in tetraploid potato. Plant Cell Rep 1–16

Venuprasad R, Dalid C, del Valle M, Zhao D, Espiritu M, Cruz MS, Amante M, Kumar A, Atlin G (2009) Identification and characterization of large-effect quantitative trait loci for grain yield under lowland drought stress in rice using bulk-segregant analysis. Theor Appl Genet 120:177–190

Venuprasad R, Bool M, Quiatchon L, Cruz MS, Amante M, Atlin G (2012) A large-effect QTL for rice grain yield under upland drought stress on chromosome 1. Mol Breed 30:535–547

Wallis JG, Wang H, Guerra DJ (1997) Expression of a synthetic antifreeze protein in potato reduces electrolyte release at freezing temperatures. Plant Mol Biol 35:323–330

Wang L, Chen L, Li R, Zhao R, Yang M, Sheng J, Shen L (2017) Reduced drought tolerance by CRISPR/Cas9-mediated SlMAPK3 mutagenesis in tomato plants. J Agric Food Chem 65:8674–8682

Winicov I, Bastola DR (1999) Transgenic overexpression of the transcription FactorAlfin1 enhances expression of the endogenous MsPRP2Gene in Alfalfa and improves salinity tolerance of the plants. Plant Physiol 120:473–480

Woo H-J, Sohn S-I, Shin K-S, Kim J-K, Kim B-G, Lim M-H (2014) Expression of tobacco tocopherol cyclase in rice regulates antioxidative defense and drought tolerance. Plant Cell Tissue Organ Cult (PCTOC) 119:257–267

Wu X, Shiroto Y, Kishitani S, Ito Y, Toriyama K (2009) Enhanced heat and drought tolerance in transgenic rice seedlings overexpressing OsWRKY11 under the control of HSP101 promoter. Plant Cell Rep 28:21–30

Xie K, Yang Y (2013) RNA-guided genome editing in plants using a CRISPR–Cas system. Mol Plant 6:1975–1983

Xu K, Mackill DJ (1996) A major locus for submergence tolerance mapped on rice chromosome 9. Mol Breed 2:219–224

Xu D-Q, Huang J, Guo S-Q, Yang X, Bao Y-M, Tang H-J, Zhang H-S (2008) Overexpression of a TFIIIA-type zinc finger protein gene ZFP252 enhances drought and salt tolerance in rice (Oryza sativa L.). FEBS Lett 582:1037–1043

Xu K, Chen S, Li T, Ma X, Liang X, Ding X, Liu H, Luo L (2015) OsGRAS23, a rice GRAS transcription factor gene, is involved in drought stress response through regulating expression of stress-responsive genes. BMC Plant Biol 15:141

Yadav SK (2010) Cold stress tolerance mechanisms in plants. A review. Agron Sustain Dev 30:515–527

Yang X, Wang X, Ji L, Yi Z, Fu C, Ran J, Hu R, Zhou G (2015) Overexpression of a Miscanthus lutarioriparius NAC gene MlNAC5 confers enhanced drought and cold tolerance in Arabidopsis. Plant Cell Rep 34:943–958

Yeo E-T, Kwon H-B, Han S-E, Lee J-T, Ryu J-C, Byu M (2000) Genetic engineering of drought resistant potato plants by introduction of the trehalose-6-phosphate synthase (TPS1) gene from Saccharomyces cerevisiae. Mol Cells 10:263–268

Ye C, Tenorio FA, Redoña ED, Morales–Cortezano PS, Cabrega GA, Jagadish KS, Gregorio GB (2015) Fine-mapping and validating qHTSF4. 1 to increase spikelet fertility under heat stress at flowering in rice. Theor Appl Genet 128:1507–1517

Yokotani N, Ichikawa T, Kondou Y, Matsui M, Hirochika H, Iwabuchi M, Oda K (2008) Expression of rice heat stress transcription factor OsHsfA2e enhances tolerance to environmental stresses in transgenic Arabidopsis. Planta 227:957–967

Yu J, Ge H, Wang X, Tang R, Wang Y, Zhao F, Lan W, Luan S, Yang L (2017) Overexpression of pyrabactin resistance-like abscisic acid receptors enhances drought, osmotic, and cold tolerance in transgenic poplars. Front Plant Sci 8:1752

Zhang H-X, Hodson JN, Williams JP, Blumwald E (2001) Engineering salt-tolerant Brassica plants: characterization of yield and seed oil quality in transgenic plants with increased vacuolar sodium accumulation. Proc Natl Acad Sci 98:12832–12836

Zhang X, Zhou S, Fu Y, Su Z, Wang X, Sun C (2006) Identification of a drought tolerant introgression line derived from Dongxiang common wild rice (O. rufipogon Griff.). Plant Mol Biol 62:247–259

Zhang H, Zhang J, Wei P, Zhang B, Gou F, Feng Z, Mao Y, Yang L, Zhang H, Xu N (2014) The CRISPR/C as9 system produces specific and homozygous targeted gene editing in rice in one generation. Plant Biotechnol J 12:797–807

Zhang C, Wohlhueter R, Zhang H (2016) Genetically modified foods: a critical review of their promise and problems. Food Sci Hum Wellness 5:116–123

Zhang A, Liu Y, Wang F, Li T, Chen Z, Kong D, Bi J, Zhang F, Luo X, Wang J (2019a) Enhanced rice salinity tolerance via CRISPR/Cas9-targeted mutagenesis of the OsRR22 gene. Mol Breed 39:47

Zhang H-X, Zhang Y, Yin H (2019b) Genome editing with mRNA encoding ZFN, TALEN and Cas9. Mol Ther

Zhao C, Zhu J-K (2016) The broad roles of CBF genes: from development to abiotic stress. Plant Signal Behav 11:e1215794

Zhao Y, Zhang C, Liu W, Gao W, Liu C, Song G, Li W-X, Mao L, Chen B, Xu Y (2016) An alternative strategy for targeted gene replacement in plants using a dual-sgRNA/Cas9 design. Sci Rep 6:23890

Zhao C, Wang P, Si T, Hsu C-C, Wang L, Zayed O, Yu Z, Zhu Y, Dong J, Tao WA (2017) MAP kinase cascades regulate the cold response by modulating ICE1 protein stability. Dev Cell 43(618–629):e5

Zhao D, Xia X, Su J, Wei M, Wu Y, Tao J (2019) Overexpression of herbaceous peony HSP70 confers high temperature tolerance. BMC Genomics 20:70

Zhu B, Chen TH, Li PH (1996) Analysis of late-blight disease resistance and freezing tolerance in transgenic potato plants expressing sense and antisense genes for an osmotin-like protein. Planta 198:70–77

Zhu B, Su J, Chang M, Verma DPS, Fan Y-L, Wu R (1998) Overexpression of a Δ1-pyrroline-5-carboxylate synthetase gene and analysis of tolerance to water-and salt-stress in transgenic rice. Plant Sci 139:41–48

Zhu YL, Pilon-Smits EA, Jouanin L, Terry N (1999) Overexpression of glutathione synthetase in Indian mustard enhances cadmium accumulation and tolerance. Plant Physiol 119:73–80

Zuo Z-F, Kang H-G, Park M-Y, Jeong H, Sun H-J, Yang D-H, Lee Y-E, Song P-S, Lee H-Y (2019) Overexpression of ICE1, a regulator of cold-induced transcriptome, confers cold tolerance to transgenic zoysia japonica. J Plant Biol 62:137–146

Chapter 6
Vegetable Crop Improvement Using CRISPR/Cas9

Francisco F. Nunez de Caceres Gonzalez and Daniela De la Mora Franco

Abstract With the human population steadily growing, the need for food security is becoming a global concern. The use of breakthrough technologies to increase the speed, quantity and quality of the vegetables we produce has become a pressing need. CRISPR/Cas9-mediated genome editing has the potential to become a game changer regarding the way crop improvement strategies will be defined in the near future. Vegetable crops are one of the main constituents of the food supply chain in the world. Due to their diversity, flavour profile and nutritional content, they have become staple foods in many regions around the globe. However, because of their physiology, they are more susceptible to damage because of climate change and several types of biotic stress. Thus, the need to obtain new varieties with improved yields, better adaptability and resistance to biotic and abiotic stress is pressing. Conventional breeding relies on the variability present in the current genetic pool, additionally, it is laborious and could take almost 20 years to produce a new commercial variety depending on the species. It is here where technologies such as CRISPR/Cas9 with its versatility, ease of use and high efficiency can yield better results. Contingent on effective implementation schemes, adequate legislation, intellectual property control and a steady investment in research, this technology could lead to a new era in vegetable crop improvement.

Keywords Vegetables · CRISPR/Cas9 · Gene editing · Leafy

6.1 Introduction

One of the most difficult challenges that the society is facing is food security. It has been estimated that by 2050, the human population will reach 10 billion and thus the global level of food production will need to be increased by approximately

F. F. Nunez de Caceres Gonzalez (✉)
Bayer Vegetable Seeds, Almeria, Spain
e-mail: francisco.nunezdecaceresgonzalez@bayer.com

D. De la Mora Franco
CINVESTAV, Irapuato, Mexico

A. Bhattacharya et al. (eds.), *CRISPR/Cas Genome Editing*, Concepts and Strategies in Plant Sciences, https://doi.org/10.1007/978-3-030-42022-2_6

60–100% (Jaganathan et al. 2018). This problem presents a difficult scenario if we also consider other factors affecting farmers and producers around the world such as climate change, increased biotic and abiotic stresses, lower availability of land, economic disparity and political and regulatory differences between countries (Ma et al. 2016). Given the difficulty or even null possibility to control many of those factors, agriculture must then rely on technological developments that can help increase not only the quantity of produce but also the quality. Due to their diversity, vegetable crops represent one of the most important food sources. They are grown worldwide and can provide vitamins, minerals, fibre and metabolites that are essential to human nutrition. On the other hand, because of their physiology, vegetables are more likely to be damaged by unfavourable environmental conditions and climate change (Karkute et al. 2017). Therefore, it is imperative to obtain new varieties that can be quickly adapted to these shifting conditions.

Traditional breeding has been important to support the constant need for higher yields for many years. Nonetheless, this method is dependent on the natural genetic variability. Additionally, in some vegetable species that genetic base has been declining due to the extensive selection, limiting the availability of alleles for further improvement (Karkute et al. 2017). Conventional breeding is also laborious and time-consuming taking up to 20 years for a new variety to reach commercial production on some species. This clearly represents a major obstacle when the need for improved cultivars is pressing. The induction of haploidy was known since the 50s but the actual implementation of double haploids in breeding programs in the 90s was a major breakthrough that has helped to vastly increase the speed to obtain new varieties (Kelliher et al. 2017). However, the dependence on the existing genetic pool is not resolved by this approach. Additionally, efficient systems to produce double haploids in important vegetable crops are scarce.

An alternative to overcome the aforementioned limitations is the use of genetic manipulation. Several techniques to introduce foreign genetic material into vegetable cells, including particle bombardment, *Agrobacterium* and direct DNA uptake, have been used for the past 30 years (Jaganathan et al. 2018). Such techniques have been crucial in understanding the biological function of genes and thus have been an excellent source for creating new traits in vegetable crops. Anyhow, the random and sometimes unstable insertion of the transgenes and recalcitrance of some of the most important vegetable species has limited its application. Moreover, the integration of unwanted genetic material coming from vectors and the use of marker genes has been a matter of public concern leading to very strict regulation worldwide (Waltz 2018). For more than a decade, techniques based on site-specific nucleases producing well defined and directed mutations have been successfully used to modify several plant species. These genome editing tools have become a powerful alternative for vegetable crop improvement. Clustered Regulatory Interspaced Short Palindromic Repeat Associated Protein System (CRISPR/Cas9) has become the preferred method compared to others Zinc Finger Nucleases (ZFNs) and Transcriptional Activator-Like Effector Nucleases (TALENs) due to its simplicity, versatility and higher frequencies (Zhang et al. 2017). In this chapter, the most recent applications of this technology on vegetable crops and the future perspectives are reviewed.

6.2 Fruit Vegetables

Tomato. The vegetable crop in which the majority of research regarding CRISPR/Cas has been done is tomato. This is a reflection of the economic importance of this species as well as the extensive genomic knowledge. Additionally, the availability of highly efficient transformation methods makes it an ideal target for gene editing. In fact, the first report for any vegetable crop using the CRISPR/Cas9 system was published by Ron et al. (2014). They used *A. rhizogenes* to transform roots of a stable transgenic line expressing *mGFP* driven by the *SCARECROW* (*SCR*) promoter from *S. lycopersicum*. The first construct was aimed at testing the system by editing the coding region of the *mGFP* gene. Roots were obtained with varying levels of gfp expression and latter analysis by restriction enzyme and PCR confirmed the edition events. Their second vector assessed the level of specificity by designing a 19 bp sgRNA homologous to *eGFP* but with only 4 nucleotides different from the *mGFP* sequence. No roots with reduced levels of GFP expression were found confirming that the sgRNA was highly specific. Finally, they targeted the gene *SHORT-ROOT* (*SlSHR*) to determine if its function was conserved among species. Their results showed a similar phenotype than the one from *Arabidopsis*, thus validating the system to perform gene knockout for functional characterisation. Later that year, another report was published related to the proof of concept of the system (Brooks et al. 2014). This group selected a gene *SLARGONAUTE* (*SLAGO7*) that when disrupted it would produce a very distinctive phenotype. The results showed T0 plants with severe presence of needle-like leaves. Such results proved that this gene edition system was again successful.

Once the first reports proved the concept of using CRISPR/Cas 9 to edit the genome of tomato, research has been performed in a broad spectrum of topics. Modification of several genes involved in developmental pathways have been edited; In 2015 Ito et al. targeted the gene *RIN* and obtained plants with incomplete ripening of the fruit; *CLV3* and *SPG5*, involved in meristematic proliferation and flowering repression respectively, were edited creating lines with increased fruit size due to abnormal growth of meristems and plants showing early flowering originated by their loss of sensitivity to long days (Xu et al. 2015; Soyk et al. 2017). Two reports on parthenocarpy have been published, Klap et al. (2017) generated mutations in the gene *SlAGL6*, resulting in plants developing seedless fruit. In the second report, the knockout of *SlIAA9* produced plants with modified leaf shape and the expected seedless fruit as well. Regarding important agronomical traits, successful generation of mutated lines have been reported with increased lycopene content (Li et al. 2018c), higher levels of aminobutyric acid (Li et al. 2018a) and a prolonged shelf life (Yu et al. 2017).

Different types of biotic and abiotic stresses have been also the focus of research in tomato. The gene *SlMAPK3* was mutated by Wang et al. (2017), the obtained lines showed lower tolerance to drought stress due to the knockout of the gene. These results determining the conserved function could prove useful for breeding purposes by selecting individuals with high expression of this gene. Plant pathogens and susceptibility of some elite cultivars to them are a major concern needed to be addressed; in this regard, the gene *SlMlo1* was the subject of knockout with the aim of conferring resistance to powdery mildew in tomato. The results showed that this approach was successful in obtaining lines that were resistant to the pathogen (Nekrasov et al. 2017). Similarly, Ortigosa et al. (2019) developed a tomato variety (Moneymaker) with resistance to bacterial speck disease caused by *Pseudomonas syringae* pv tomato. They focused on mutating the *SlJAZ2* gene, generated truncated JAZ2 forms lacking the C-terminal Jas domain. This edited variety showed a reduced bacterial entry through the stomata but not compromising its resistance to necrotrophs. In addition to targeting desirable phenotypic and agronomical traits, researchers have also looked for new applications to this technology. In 2017 Jacobs et al. proposed the use of pooled CRISPR/Cas9 libraries to generate mutated populations as an alternative to other mutagenesis methods such as EMS or fast neutron. Their results showed high specificity on the targeted genes but also high efficiency in generating such mutations. Other reports have been published on phenotypical traits, biotic and abiotic stress, nutritional value and plant architecture. Table 6.1 presents an updated list of the current reports on the use of CRISPR/Cas9 system in tomato and other vegetable crops.

Cucumber. Within the *Curcubitaceae* family, cucumber was the first species with a successful report of gene editing by the CRISPR-Cas9 system (Chandrasekaran et al. 2016). The authors targeted the *elF4E* gene at two different sites for the development of broad virus resistance. They succeeded in conferring broad viral resistance to cucumber but the efficiency of the procedure was very low. Hu et al. (2017) focused on producing gynoecious lines through CRISPR-Cas9 mediated mutagenesis of the *CsWIP1* gene. The system was optimised by using the stronger CsU6 promoter. T0 mutants showed a gynoecious phenotype, with the upper nodes carrying only female flowers and smaller leaves compared to the wild type plants.

Watermelon. This species was the subject of gene editing in 2016 by Tian et al. who aimed at creating knockout mutations in the *ClPDS* gene. Disruption was successful and it originated plantlets with the expected albino phenotype but the generation of weak shoots gradually proved to be lethal for these lines. The *Acetolactate synthase (ALS)* gene in watermelon was the target for editing in a later study with the aim to obtain engineered herbicide-resistant plants (Tian et al. 2018). Interestingly, a point mutation was attempted in this study. The change form C to T in one of the codons of this gene could confer herbicide resistance. They developed 199 plants of which, 45 had the expected mutation from C to T. Then the homozygous lines were treated with the herbicide tribenuron, and 14 days after the treatment these edited plants, showed no damage, whatsoever, compared to the highly affected WT plants.

Table 6.1 Current landscape of CRISPR/Cas9 studies in vegetable crops

Crop	Target gene	Trait	References
Carrot	*F3H*	Phenotypic	Klimek et al. (2018)
Chicory	*CIPDS*	Phenotypic	Bernard et al. (2019)
Cucumber	*elF4E* *CsWIP1*	Virus resistance Gynoecious induction	Chandrasekaran et al. (2016) Hu et al. (2017)
Lettuce	*BIN2* *LsNCED4*	Developmental Abiotic stress	Woo et al. (2015) Bertier et al. (2018)
Potato	*StALS1* *StIAA2* *StALS1* *StMYB44* *GBSS*	Herbicide resistance Developmental Herbicide resistance Abiotic stress Starch quality	Butler et al. (2015) Wang et al. (2015) Butler et al. (2016) Zhou et al. (2017) Andersson et al. (2017, 2018)
Tomato	*mGFP, SlSHR* *SLAGO7* *RIN* *ANT1* *CLV3* *SlPDS, SlPIF4* *SPG5* *PSY1* *ALC* *LRR-XII* family, *PDS* *SlAGL6, SlIAA9* *DELLA, ETR1* *SlMlo1* *SlMAPK3* *SlGAD2, SlGAD3* *SlIAA9* *GABATP1-3, CAT9,* *SSADH* *SLCBF1* *SGR1, LCY-E, Blc,* *LCY-B, LCY-B2* *CRITISO, PSY1* *SLMYB12* *SLJAZ2* *DELLA family*	Gene functional characterisation Gene functional characterisation Fruit ripening Phenotypic Fruit size Phenotypic Early flowering Phenotypic Prolonged shelf life Phenotypic/Mutagenised populations Parthenocarpy Developmental Fungal resistance Drought tolerance GABA increase Parthenocarpy control Increase in aminobutyric acid Abiotic stress Lycopene content Phenotypic Phenotypic Bacterial resistance Plant architecture	Ron et al. (2014) Brooks et al. (2014) Ito et al. (2015) Čermák et al. (2015) Xu et al. (2015) Pan et al. (2016) Soyk et al. (2017) Hayut et al. (2017) Yu et al. (2017) Jacobs et al. (2017) Klap et al. (2017) Shimatani et al. (2017) Nekrasov et al. (2017) Wang et al. (2017) Nonaka et al. 2017 Ueta et al. (2017) Li et al. (2018a) Li et al. (2018b, c) Dahan et al. (2018) Deng et al. (2018) Ortigosa et al. (2019) Tomlinson et al. (2019)
Watermelon	*ClPDS* *ALS*	Phenotypic Herbicide resistance	Tian et al. (2016) Tian et al. (2018)

6.3 Root and Bulb Vegetables

Potato. It has become a very important crop worldwide and it is now considered a staple crop in Europe and many parts of the Americas. The need for varieties that can adapt to climate change and to improve their nutritional value is the higher priority in the current breeding programs. The use of gene edition technology represents an excellent opportunity for this crop. The CRISPR/Cas9 system was reported for the first time in potato by Butler et al. (2015). They used diploid and tetraploid varieties to modify the *StALS1* gene. Somatic mutations were most evident in the diploid background with three of the four primary events having more than two mutation types at a single *ALS* locus. Conversely, the tetraploid background, had only one mutation type in four of the five candidates. Single targeted mutations were inherited through the germline of both backgrounds with transmission percentages ranging from 87 to 100%. Wang et al. (2015), used CRISPR-Cas9 to produce knock-outs in the *StIAA2* gene in potato which is involved in shoot morphogenesis. They obtained both homozygous and heterozygous plants with different mutations, proving that the system was successful in producing directed mutations for gene functional characterisation. The *ACETOLACTATE SHYNTHASEI (ALSI)* gene was targeted for modification to confer reduced susceptibility to ALS-inhibiting herbicides in potato (Butler et al. 2016). Their approach was to obtain edited events based on homologous recombination and thus they included a repair template in their vector. They also used two methods for transformation, *Agrobacterium* and a Geminivirus Replicon (GVR) system. Interestingly, their results showed that only the plants obtained by the GVR method acquired an increased herbicide tolerance. Nutritional content was targeted by Andersson et al. (2017) to increase the starch quality by knocking out the *GBSS* gene. They obtained yielded mutations in all four alleles in a single transfection, in up to 2% of regenerated lines. Based on these results, Andersson et al. (2018) targeted the same gene but implementing this time a DNA-free genome editing method using delivery of CRISPR-Cas9 ribonucleoproteins (RNPs) to potato protoplasts. RNA induced mutations were found at a frequency of up to 9% with all mutated lines being transgene-free. However, more than 80% of the shoots with confirmed mutations had unintended inserts in the cut site, which was within the range compared to DNA delivery.

Carrot. One of the most highly-valuable vegetable crops and it has been used as a model species in biotechnology for decades. Utilised for the development of plant tissue culture techniques and also amenable to genetic transformation using *Agrobacterium*. However, research progress in carrot has been much slower than in other crop species mainly due to biennial reproductive cycle, out-crossing, and high inbreeding depression effect. Nevertheless, Klimek et al. in 2018 reported their results on gene editing in this species using the CRISPR-Cas9 system for targeted mutagenesis. Multiplexing CRISPR-Cas9 vectors expressing two single-guide RNA (gRNAs) targeting the carrot flavanone-3-hydroxylase *(F3H)* gene were tested for the blockage of the anthocyanin biosynthesis. The study was also aimed at evaluating the editing efficiency of three codon-optimised SpCas9 genes (AteCas9, zCas9 and Cas9p) to identify the most efficient system for the carrot. The obtained results showed that the AteCas9 system was the most effective in generating targeted mutations reaching up to an impressive 90% of efficiency. Knockout of *F3H* gene resulted in the discolouration of calli, validating the functional role of this gene in the anthocyanin biosynthesis pathway.

6.4 Leafy Vegetables

Lettuce. It is considered a major crop worldwide with a short life cycle and its production represents a multibillion-dollar industry. The wealth on genomic information on this species makes it a good study subject for genome editing. Research being done on this species regarding the CRISPR-Cas system was first published by Woo et al. (2015). They reported the direct delivery of ribonucleoproteins (RNPs) into protoplast cells and the induction of targeted genome modifications in various plant species including lettuce. They transfected protoplasts to induce mutations in the *BIN2* gene, which encodes a negative regulator in the brassinosteroid signalling pathway. Whole plants were successfully regenerated and later analysis revealed that they were able to inherit the mutated allele to their progeny. Bertier et al. (2018), targeted the *LsNCED4* gene which regulates thermo-inhibition of seed germination in lettuce. Analysis of 47 primary transformants (T_1) and 368 T_2 showed that knockout of this gene resulted in large increases in the maximum temperature tolerance for seed germination, with more than 70% of seeds capable of germinating at 37 °C.

Chicory. Beyond being an important vegetable crop, it is a species well known for its production of secondary metabolites with potential medicinal use and also utilized in the food industry for inulin production which is a sugar substitute. This has contributed to large-scale cultivation and therefore, to an interest in genetically modified chicory to improve yield and nutritional properties. Bernard et al. (2019) reported CRISPR/Cas9-mediated targeted mutagenesis in chicory. They selected the *CiPDS*, expecting to obtain an evident albino phenotype. As a result, using *Agrobacterium* and the direct delivery in protoplasts, they were able to recover albino plants at frequencies ranging 4.5–31.3% depending on the method. Such results suggest that this system could lead to more research aimed at increasing the nutraceutical value of this crop in the future.

6.5 Discussion

With the ever-growing need for food supply in the world, agriculture has to rely on breakthrough technologies to overcome the many difficulties that genome editing represents. The use of conventional breeding is laborious and slow in producing better varieties to reach the market. Other alternatives, such as transgenesis have low acceptance by the public and thus their application has become very limited (Gupta and Musunuru 2014). Considering this, sequence-specific nucleases are becoming a very powerful tool for crop improvement. TALENs and ZFNs were promising alternatives in terms of gene edition. However, they had limitations in their design and applications. The accessibility, simplicity and precision of the CRISPR/Cas9-mediated mutagenesis have now provided a much better option that is revolutionising the way crop improvement is going to be perceived in the near future (Xie et al. 2013; Arora and Narula 2017). Moreover, this system is so versatile that can be used for different applications such as gene knockout, point mutations, mutagenised libraries, or even insertion of templates by homologous recombination (Bortesi and Fischer 2015). Although the CRISPR/Cas9 system is an excellent tool for genome editing, it also has some disadvantages like off-target mutations due to perceived low specificity. Some of the Cas9 proteins are not equally efficient in all the crops that have been tested, limiting its application (Belhaj et al. 2015). It is also very important to consider in the development of this technology, the regulation and government policy regarding the commercialisation of gene-edited plants. The ruling of the EU in this matter, for example has had a strong impact not only on the amount of research being done in that part of the world but also in the public perception (CURIA 2018).

Another key factor in the future success of this system will be the method of delivery to the plant cells. So far, the majority of reports have used well-established transformation protocols based on *Agrobacterium*. This stable integration can have two major outcomes also related to GMOs, the first and more important being the probability of vector DNA being also transferred to the host genome permanently. The other one is the use of selection genes in the vector construction. Permanent integration can also produce a higher rate of off-target cleavage mutations if the sgRNA was not properly designed or has high homology with other genomic regions. Thus, the need for transient but effective delivery methods gains a huge relevance. Development of transfection protocols into single cells such as magnetofection, lipofection or sonoporation are some examples of technologies that researchers are looking into to avoid the use of methods closely associated with GMOs.

To date, not many vegetable crops have been targeted for genome editing using the CRISPR system, tomato and potato account for almost 80% of the reports. This can be attributed to the importance of both crops, being staple foods in several parts of the world. Additionally, genetic transformation, molecule delivery and regeneration methods are readily available for these species and thus these makes them a natural choice for this type of research (Jaganathan et al. 2018). On the other hand, other important vegetable crops such as pepper and onion are lagging behind because of their recalcitrant nature for genetic manipulation. Thus, the development of methods for efficient regeneration and/or molecule delivery will be required for these and other crops if genome edition is going to be a reality for them. Fortunately, it has been proved that direct delivery of ribonucleoproteins into cells is effective to achieve mutations (Andersson et al. 2018), therefore this approach could be an excellent alternative to avoid the necessity of in vitro regeneration.

It is undeniable that the CRISPR/Cas9-mediated genome editing can modify how vegetable crop improvement strategies will be defined both in the academical and the commercial areas in the near future. We strongly believe that if the right implementation schemes, appropriate legislation, intellectual property control and continuous investment come together this could lead to a new era in food production.

References

Andersson M, Turesson H, Olsson N et al (2018) Genome editing in potato via CRISPR-Cas9 ribonucleoprotein delivery. Physiol Plantarum 164:378–384

Andersson M, Turesson H, Nicolia A et al (2017) Efficient targeted multiallelic mutagenesis in tetraploid potato (*Solanum tuberosum*) by transient CRISPR-Cas9 expression in protoplasts. Mol Plant 36:117–128

Arora L, Narula A (2017) Gene editing and crop improvement using CRISPR-Cas9 system. Front Plant Sci 8:1932

Bortesi L, Fischer R (2015) The CRISPR/Cas9 system for plant genome editing and beyond. Biotechnol Adv 33:41–52

Belhaj K, Chaparro A, Kamoun S et al (2015) Editing plant genomes with CRISPR/Cas9. Curr Opin Biotech 32:76–84

Bernard G, Gagneul D, Alves Dos Santos H et al (2019) Efficient genome editing using CRISPR/Cas9 technology in chicory. Int J Mol Sci 20:1155

Bertier LD, Ron M, Huo H et al (2018) High-resolution analysis of the efficiency, heritability, and editing outcomes of CRISPR/Cas9-induced modifications of *NCED4* in lettuce (*Lactuca sativa*). G3 8:1513

Brooks C, Nekrasov V, Lippman ZB et al (2014) Efficient gene editing in tomato in the first generation using the clustered regularly interspaced short palindromic repeats/CRISPR-Associated9 system. Plant Physiol 166:1292–1297

Butler NM, Atkins PA, Voytas DF et al (2015) Generation and inheritance of targeted mutations in potato (*Solanum tuberosum* L.) using the CRISPR/Cas system. PLoS ONE 10(12): e0144591. https://doi.org/10.1371/journal.pone.0144591

Butler NM, Baltes NJ, Voytas DF et al (2016) Geminivirus-mediated genome editing in potato (*Solanum tuberosum* L.) using sequence-specific nucleases. Front Plant Sci 7:1045

Čermák T, Baltes NJ, Čegan R et al (2015) High-frequency, precise modification of the tomato genome. Genome Biol 16:232

Chandrasekaran J, Brumin M, Wolf D et al (2016) Development of broad virus resistance in non-transgenic cucumber using CRISPR/Cas9 technology. Mol Plant Pathol 17:1140–1153

CURIA Court of Justice of the European Union (2018) https://curia.europa.eu/jcms/upload/docs/application/pdf/2018-07/cp180111en.pdf. Accessed 7 Jan 2019

Dahan T, Filler S, Melamed C et al (2018) Efficient *in planta* gene targeting in tomato using geminiviral replicons and the CRISPR/Cas9 system. Plant J 95:5–16

Deng L, Wang H, Sun C et al (2018) Efficient generation of pink-fruited tomatoes using CRISPR/Cas9 system. J Genet Genomics 45:51–54

Gupta RM, Musunuru K (2014) Expanding the genetic editing tool kit: ZFNs, TALENs, and CRISPR-Cas9. J Clin Invest 124:4154–4161

Hayut SF, Bessudo CM, Levy AA (2017) Targeted recombination between homologous chromosomes for precise breeding in tomato. Nat Commun 8:15605

Hu B, Li D, Liu X et al (2017) Engineering non-transgenic gynoecious cucumber using an improved transformation protocol and optimized CRISPR/Cas9 system. Mol Plant 10:1575–1578

Ito Y, Nishizawa A, Endo M et al (2015) CRISPR/Cas9-mediated mutagenesis of the *RIN* locus that regulates tomato fruit ripening. Biochem Bioph Res Co 467:76–82

Jacobs TB, Zhang N, Patel D et al (2017) Generation of a collection of mutant tomato lines using pooled CRISPR libraries. Plant Pysiol 174:2023–2037

Jaganathan D, Ramasamy K, Sellamuthu G et al (2018) CRISPR for crop improvement: an update review. Front Plant Sci 9:985

Karkute SG, Singh AK, Gupta OP et al (2017) CRISPR/Cas9 mediated genome engineering for improvement of horticultural crops. Front Plant Sci 8:1635

Kelliher T, Starr D, Richbourg L et al (2017) MATRILINEAL, a sperm-specific phospholipase, triggers maize haploid induction. Nature 542:105–109

Klap C, Yeshayahou E, Bolger AM et al (2017) Tomato facultative parthenocarpy results from Sl*AGAMOUS-LIKE 6* loss of function. Plant Biotechnol J 15:634–647

Klimek M, Oleszkiewicz T, Lowder LG et al (2018) Efficient CRISPR/Cas9-based genome editing in carrot cells. Plant Cell Rep 37:575–586

Li R, Li R, Li X et al (2018a) Multiplexed CRISPR/Cas9-mediated metabolic engineering of c-aminobutyric acid levels in *Solanum lycopersicum*. Plant Biotechnol J 16:415–427

Li R, Zhang L, Wang L et al (2018b) Reduction of tomato-plant chilling tolerance by CRISPR−Cas9-mediated *SlCBF1* mutagenesis. J Agric Food Chem 66:9042–9051

Li X, Wang Y, Chen S et al (2018c) *Lycopene* is enriched in tomato fruit by CRISPR/Cas9-mediated multiplex genome editing. Front Plant Sci 9:1–2

Nekrasov V, Wang C, Win J et al (2017) Rapid generation of a transgene-free powdery mildew resistant tomato by genome deletion. Sci Rep UK 7:1–6

Nonaka S, Arai C, Takayama M et al (2017) Efficient increase of ɣ-aminobutyric acid (GABA) content in tomato fruits by targeted mutagenesis. Sci Rep UK 7:7057

Ma X, Zhu Q, Chen Y et al (2016) CRISPR/Cas9 platforms for genome editing in plants: developments and applications. Mol. Plant 9:961–974

Ortigosa A, Gimenez S, Leonhardt N et al (2019) Design of a bacterial speck resistant tomato by CRISPR/Cas9-mediated editing of *SlJAZ2*. Plant Biotechnol J 17:665–673

Pan C, Ye L, Qin L et al (2016) CRISPR/Cas9-mediated efficient and heritable targeted mutagenesis in tomato plants in the first and later generations. Sci Rep UK 6:24765

Ron M, Kajala K, Pauluzzi G et al (2014) Hairy root transformation using *Agrobacterium rhizogenes* as a tool for exploring cell type-specific gene expression and function using tomato as a model. Plant Physiol 166:455–469

Shimatani Z, Kashojiya S, Takayama M et al (2017) Targeted base editing in rice and tomato using a CRISPR-Cas9 cytidine deaminase fusion. Nat Biotechnol 35:441–443

Soyk S, Müller NA, Ju Park S et al (2017) Variation in the flowering gene *SELF PRUNING 5G* promotes day-neutrality and early yield in tomato. Nat Genet 49:162–168

Tian S, Jiang L, Cui X et al (2018) Engineering herbicide-resistant watermelon variety through CRISPR/Cas9-mediated base-editing. Plant Cell Rep 37:1353–1356

Tian S, Jiang L, Gao Q et al (2016) Efficient CRISPR/Cas9-based gene knockout in watermelon. Plant Cell Rep 36:399–406

Tomlinson L, Yang Y, Emenecker R et al (2019) Using CRISPR/Cas9 genome editing in tomato to create a gibberellin-responsive dominant dwarf DELLA allele. Plant Biotechnol J 17:132–140

Ueta R, Abe C, Watanabe T et al (2017) Rapid breeding of parthenocarpic tomato plants using CRISPR/Cas9. Sci Rep UK 7:507

Waltz E (2018) With a free pass, CRISPR-edited plants reach market in record time. Nat Biotechnol 36:6–7

Wang L, Chen L, Li R et al (2017) Reduced drought tolerance by CRISPR/Cas9-mediated Sl*MAPK3* mutagenesis in tomato plants. J Agric Food Chem 65:8674–8682

Wang S, Zhang S, Wang W et al (2015) Efficient targeted mutagenesis in potato by the CRISPR/Cas9 system. Plant Cell Rep 34:1473–1476

Woo JW, Kim J, Kwon SI et al (2015) DNA-free genome editing in plants with preassembled CRISPR-Cas9 ribonucleoproteins. Nat Biotechnol 33:1162–1164

Xie K, Yang Y (2013) RNA-guided genome editing in plants using a CRISPR–Cas system. Mol Plant 6:1975–1983

Xu C, Liberatore KL, MacAlister CA et al (2015) A cascade of arabinosyltransferases controls shoot meristem size in tomato. Nat Genet 47:784–792

Yu Q, Wang B, Li N et al (2017) CRISPR/Cas9-induced targeted mutagenesis and gene replacement to generate long-shelf life tomato lines. Sci Rep UK 7:11874

Zhang H, Zhang J, Lang Z et al (2017) Genome editing —principles and applications for functional genomics research and crop improvement. Crit Rev Plant Sci 36:291–309

Zhou X, Zha M, Huang J et al (2017) StMYB44 negatively regulates phosphate transport by suppressing expression of *PHOSPHATE1* in potato. J Exp Bot 68:1265–1281

Chapter 7
Use of CRISPR in Climate Smart/Resilient Agriculture

Vinod Kumar, Sabah AlMomin, Muhammad Hafizur Rahman, and Anisha Shajan

Abstract Recent developments in genome editing have opened new vistas for functional genomics and crop improvement. The CRISPR (**C**lustered **R**egularly **I**nterspaced **S**hort **P**alindromic **R**epeat)/Cas9 system is a powerful tool for genome editing. The exponentially growing world population will need more food with better nutritional content, improved stress tolerance, enhanced disease resistance, and there is no other way to ignore the benefits of advanced crop improvement tools such as genome editing for climate-smart sustainable agriculture. The discovery of high precision genome-editing tools has advanced plant genetic engineering to newer heights. CRISPR/Cas9 was adopted from a naturally occurring genome editing system in bacteria. Unlike the previous generation genome editing tools such as zinc-finger nucleases (ZFNs) and transcription activator-like effector nucleases (TALENs), CRISPR/Cas9 brings in simplicity in cloning, versatility in deploying various guide RNAs, and cost-effectiveness. A number of crop species have been successfully subjected to genome editing and this technique shows great potential towards achieving global food security. From a consumer point of view, safety aspect of biotech products is of paramount importance; however, the regulatory procedures need to evolve continuously to cope up with the rapid changes happening in the biotechnology sector. This chapter discusses the significance of this novel genome editing technique and its potential for crop improvement.

Keywords Abiotic stress · Base editors · Biotic stress · CRISPR/Cas9 · Disease resistance · Gene knockout · Genome editing · Genome engineering · Off-target mutations · Oligonucleotide-directed mutagenesis · Plant genetic engineering · Precision breeding · Sequence-specific nucleases · Small RNA · Targeted mutagenesis

V. Kumar (✉) · S. AlMomin · M. H. Rahman · A. Shajan
Biotechnology Program, Environment and Life Sciences Research Center, Kuwait Institute for Scientific Research, PO Box 24885, Safat 13109, Kuwait
e-mail: vinodk@kisr.edu.kw

A. Bhattacharya et al. (eds.), *CRISPR/Cas Genome Editing*, Concepts and Strategies in Plant Sciences, https://doi.org/10.1007/978-3-030-42022-2_7

7.1 Introduction

Biotechnological advancements have dramatically revolutionized the living world in the past three to four decades. Plant biotechnology is one such field that has helped improve the quality of agriculture worldwide. Many new techniques have been developed to produce novel crops with better traits. Since the beginning of agriculture, people have constantly worked on finding new ways for plant breeding that can improve both quality and quantity of plants and plant products. The recent development in genetic engineering techniques has taken this search one step ahead (Tourte 2019). The new breeding techniques focus on making changes on a genetic level to bring the desired changes such as total nutritional content, resistance towards diseases, endurance towards unfavorable weather conditions, etc., in plants. Apart from that, there is a huge gap between demand and supply for edible crops due to a rapidly growing population of the world (Valin et al. 2014). Next-generation genome-edited plants can be engineered to survive and grow under adverse environmental conditions, resist pests and pathogens, and contain more nutritional value than their conventional counterparts (Bao et al. 2019a, b; Bilichak et al. 2020; Schindele et al. 2020; Zhang et al. 2018a, b). Therefore, it is crucial to advance this technology with innovative discoveries that would eventually lead to the creation of crop plants with novel traits and products for the benefit of humankind.

Although there is availability of several genetic improvement tools and high throughput analysis techniques, the general public's view of foods derived from genetically-modified plants is apparently the greatest hurdle for consumer acceptance and commercialization (ISAAA 2017; Lucht 2015; Wolt et al. 2016). Hence there is a growing need to create public awareness and implement regulatory processes on scientific basis that assures the safety of new plant breeding technologies including genome editing. Despite the existing controversies, researchers and scientists are working on developing effective approaches to produce better crops with high precision in genetic modification to reduce off-target effects. Genome editing is one such technique that has become very popular in recent years due to its high precision, simplicity and efficacy (Bao et al. 2019a, b; Bilichak et al. 2020; ISAAA 2017; Schindele et al. 2020; Zhang et al. 2018a, b). In modern-day agriculture, this technique has helped in enhancing production, improving the nutritional content of crops, preventing diseases by enhancing plants' resistance or enabling plants to synthesize unique compounds through creating predicted changes in the gene sequence, or precise insertion of an exogenous DNA with the goal of inactivating gene(s), generating functional alleles, replacing mutant alleles or site-specific transgene integration (Petolino et al. 2016).

7.2 Classical Crop Improvement Techniques

Selection, crossing and other means of breeding. Historically plant selection has been used as an effective tool towards producing plants with superior traits. The vast diversity of gene pool and the presence of heterogeneous population has enabled selection of elite individuals and their propagation. Higher productivity and yield per unit area of land, improved nutritional quality and taste, extended shelf life, resistance/tolerance against biotic and abiotic stress factors have become the prime parameters of selection. This has eventually led to the development of marker-assisted selection in the era of modern-day molecular biotechnology and plant breeding. Developments in the field of plant breeding have led to the production of hybrids. Crossing of sexually compatible plants has enabled the production of hybrids that carries a combination of desirable genes from superior parents in the resultant individuals. However, because of the non-specific pooling of genes and recombinations in the crossed plants, it was necessary to create hundreds or thousands of hybrid progeny to identify those few that possess desired characteristics with a minimum of off-target effects. Although interspecific crossing was successful in a few plant species, its widespread applicability was limited to only a certain number of compatible species. Attempts were made to transfer large segments of chromosomes along with a number of neutral or detrimental genes in the past resulting in genetic (linkage) drag. However, there was limited utility of this technique due to technical challenges (Gupta and Tsuchiya 1991).

Induced variations. One of the primary requirements for selection and plant breeding is generating a population that possess variation in the gene pool. The variants are subsequently used as parents in a breeding program, to generate individuals containing a combination of favorable traits (variants) coming from contrasting parents. Somaclonal variation is nothing but the spontaneous mutations that occur when plant cells are grown in vitro most likely due relaxed control of cell division during the production of callus and plantlet differentiation. Tissue culture-induced mutations were used as a new source of genetic variability and used successfully to develop superior genotypes by plant breeders (Larkin and Scowcroft 1981). Subsequently, researchers developed more powerful means of generating variations at the genetic level by inducing mutations. Since then, mutation breeding has been used to accelerate the process of developing different traits for selection. The process involves subjecting the plants or seeds to various mutagenic agents such as ionizing radiation (e.g. γ-rays) or chemical mutagens (e.g. EMS) to induce genetic changes in the DNA. However, there was no specificity to the genetic changes, nor any means to control the effects of the mutagen on particular (target) genes or traits. Usually, there are more deleterious mutations than useful ones, making it a highly laborious task to select the ones with minimum off-target effects. In spite of all these limitations, there are more than 3,200 mutant varieties that have officially been released for commercial use comprising of over 210 plant species from more than 70 countries (Maluszynski 2001; Suprasanna et al. 2015).

7.3 Genetic Engineering and Genome Editing

Plant biologists realized the limitations of plant improvement through induced mutations and selection since the genetic changes were random, and lacked specificity when it came to trait introduction or improvement. This has led to the developments towards the introduction of specific fragments of DNA through genetic engineering tools. Genetic engineering technology allows the transfer of genes for specific traits between species using laboratory techniques. There are mainly two types of genetic engineering techniques, which are more common for plant breeding: indirect and direct. The indirect approach involves the use of a vector, such as *Agrobacterium tumefaciens*, to introduce a foreign gene into the host plant, while direct approaches are not dependent on vectors for genetic transformation. The three of the most common direct approaches are biolistic transformation, electroporation, and polyethylene glycol (PEG)-mediated protoplast transformation. In electroporation, high voltage electric power is used to deliver DNA into the host cell, and in PEG-mediated protoplast transformation, DNA is directly introduced in the protoplast of the plant. Although these techniques have shown promising outcomes in producing plant species with desired characteristics, the introduction of a whole gene sequence can also produce many unexpected and unwanted results due to their random insertion. The success of genetic engineering and transgenic methods depends on several factors, such as selection of desired sequence and right vectors, the introduction of foreign DNA into the host cell, expression of recombinant sequence, and selection and production of transformed cells. Applications of this technology include herbicide resistance, insect resistance, nutritional enhancements, stress tolerance, disease resistance, biofuel efficiency, remediation of polluted sites, etc.

For thousands of years, various plant breeding techniques have been adopted to produce crops with better genetic traits (Allard 1999; Poehlman 2013; van Harten 1998). As described in the previous sections, the plants with superior traits were selected and crossed with each other to obtain the progeny with (combined) improved qualities coming from different parents. The discovery of genes and their effect on various traits opened new avenues for crop improvement. Over the years, researchers have used chemical mutagens, irradiation, and insertion of recombinant DNA from various species to introduce desired traits in plants and animals. However, such genetic changes were not always fully accepted by general public in a few regions due to safety or ethical concerns (Bawa and Anilakumar 2013; Lucht 2015; Ricroch et al. 2018). The public concern regarding genetically-modified (GM) foods and crops apparently focused on human and environmental safety, consumer choice, and IP rights (Bawa and Anilakumar 2013). The emergence of modern-day genome editing tools can put to rest many such apprehensions. Genome editing is a type of precision genetic engineering, where in specific fragments of DNA is inserted, deleted, modified or replaced in the genome of a living organism (Kanchiswamy et al. 2015; Kim and Kim 2016; Langner et al. 2018; Mohanta et al. 2017; Sprink et al. 2015; Townsend et al. 2009; Zhang et al. 2018a, b). Genome editing is considered relatively a safer approach and the crops that have been produced using such techniques, and have therefore been proposed to be subject to separate regulatory laws than GMOs (Huang et al. 2016; Wolt et al. 2016).

7.4 History of Genetic Engineering

The recombinant DNA technology came into existence during the 1970s by producing functional plasmids by using endonuclease-generated fragments as well as by joining of plasmid—DNA molecules of entirely different origins (Cohen et al. 1973). This enabled researchers with a strikingly new way to produce organisms with desired traits. The very first GM mouse was created by microinjecting the explanted mouse blastocysts with viral DNA (Jaenisch and Mintz 1974), and the first GM plant was developed by transforming tobacco with antibiotic resistance gene through *Agrobacterium*-mediated genetic engineering (Bevan et al. 1983). The Flavr Savr™ tomato became the first commercialized FDA approved GM food that contained an antisense RNA to regulate the expression of the enzyme polygalacturonase in ripening tomato fruit (Kramer and Redenbaugh 1994). Subsequently, several plant species have been modified to create special features, such as resistance towards common herbicides or pesticides, disease resistance and tolerance towards unfavorable climatic conditions (Jones 1999; Klümper and Qaim 2014; Parekh 2004; Uzogara 2000). On the other hand, some of the plants were engineered to produce medicinal products, and increased protein or starch content of fruits and vegetables (Ma et al. 2003; Wilson and Roberts 2014; Zhu et al. 2007). In the last few years, the tremendous growth in research and development and adoption of crops resulted in massive increase in global area of biotech/GM crops reaching 189.8 million hectares in 2017 (ISAAA 2017). From 1996–2016, biotech crops provided $186.1 billion in economic gains to around 17 million farmers (ISAAA 2017).

7.5 Genome Editing

The first generation transgenic plants were created by integrating a single or a set of genes of interest (cisgene/transgene) at random locations of the genome along with marker genes for selection. Although there is a large number of peer-reviewed literature on the economic and environmental benefits of genetically engineered crops, the commercialization of GM crops faces increasing challenge due to the regulatory hurdles (Smyth 2017). This has resulted in a massive failure to adopt and commercialize a number of genetically engineered crop plants in many countries, especially in the European Union. The development of newer technologies such as oligonucleotide-directed mutagenesis, zinc finger nuclease, meganuclease technique, transcriptional activator-like effector–nuclease and CRISPR demonstrated higher degree of specificity in precise editing of DNA sequences. According to expert views, this can be classified as a form of site-directed mutagenesis in the United

States, with the resulting products not being subject to GM crop regulations (Smyth 2017). These newer forms of genome editing technique are expected to make genetic engineering-mediated plant improvement more socially acceptable. The current regulatory norms and the implications are discussed in detail from the viewpoint of public acceptance (Ishii and Araki 2017). The key feature of genome editing is the ability to change DNA sequences at precise locations with a higher degree of accuracy.

Some of the major genome editing tools used to edit plant genomes are: Homologous recombination (HR), zinc finger nucleases (ZFNs), transcription activator-like effector nucleases (TALENs), pentatrico peptide repeat proteins (PPRs), the CRISPR/Cas9 system, RNA interference (RNAi), cisgenesis, and intragenesis (Mohanta et al. 2017). A number of tools that have recently been developed to edit the genome at a single nucleotide level include site-directed sequence editing and oligonucleotide-directed mutagenesis. A detailed review of these methods has recently been published by Mohanta et al. (2017).

7.6 Genome Editing: Types and Mechanisms

Site-directed nucleases (SDNs) technology: This technique requires the use of site-directed nucleases (SDNs), a specific type of proteins, to recognize and cut a specific site on a DNA sequence. They can be categorized into SDN 1, 2 and 3 (Podevin et al. 2013). SDN1 mediated editing involves processes of non-homologous end-joining (NHEJ), an error-prone mechanism causing point mutations during repair of double-strand DNA breaks in plants. SDNs may either consist of a single chain (e.g. meganucleases) or two chains linked by a small peptide sequence (e.g. zinc finger nucleases and transcription-activator-like effector nucleases TALENs) (Lusser et al. 2011; Podevin et al. 2013). SDNs are mainly used to induce artificial mutation by deletion, insertion or replacement of a particular DNA sequence. This technique was used in producing crops with traits like increased protein or vitamin content, or resistance towards diseases. Such variations are accomplished without the introduction of a foreign DNA sequence and are the direct results of changes in plants' own biochemical processes due to deletion/replacement of a specific gene sequence. Once the specific gene sequence is detected, the nucleases can be used to create double strand breaks (DSB) (Cardi et al. 2017). After that, the plant starts its DNA repair mechanism to fill the loss, which is attained by one of the two mechanisms: either non-homologous end joining (NHEJ) or homology-directed repair (HDR) (Cardi et al. 2017). In NHEJ, small changes (insertion or deletion) are incorporated via frame-shift mutations that lead to a novel phenotype through the loss of function of the targeted gene (Araki and Ishii 2015) whereas, in the HDR mechanism, a homologous sequence is inserted to replace the target sequence. This newly inserted sequence, usually, act as a regulatory element and produce desired results by repairing the DNA.

7.6.1 *Zinc-Finger Nucleases*

Zinc-finger nucleases (ZFNs) are artificial restriction enzymes that contain zinc finger DNA-binding domain and a separate DNA-cleavage domain (Carroll 2011). Zinc finger domains can be engineered to target specific desired DNA sequences and this enables zinc-finger nucleases to target unique sequences within complex genomes. By taking advantage of endogenous DNA repair machinery, one can precisely alter the genomes of higher organisms (Urnov et al. 2010). The ZFNs can recognize both forward and reverse strands and bind on either side of the target sequence (Carroll 2011; Shukla et al. 2009). It creates DSBs in the target DNA, which is subsequently repaired by either of the above-described mechanisms (NHEJ, HDR).

7.6.2 *Transcription Activator-Like Effectors Nucleases*

Apart from ZFNs, transcription activator-like effectors nucleases (TALENs) are also used to induce DSBs. These nucleases have an affinity for at least one of the four DNA base pairs and bind on specific sites to induce mutations by insertion or deletion of a gene sequence (Bogdanove and Voytas 2011). TALENs have few advantages over ZFNs, such as they are easy to design and their length can be extended as per requirements. However, TALENs are comparatively larger nucleases, which make them unsuitable for smaller sequences. Both ZFNs and TALENs are frequently used to induce desired mutations in plants, however, none of them are devoid of disadvantages (Carroll 2011; Gupta and Musunuru 2014; Mohanta et al. 2017).

7.6.3 *CRISPR/Cas9 System*

Francisco Mojica was the first researcher to characterize a CRISPR locus in 1993 (Mojica et al. 2005). During the same time, similar observations about CRISPR elements were also made by another research group (Pourcel et al. 2005). In *Escherichia coli*, spacer sequences, which are derived from bacteriophage, are transcribed into small RNAs, termed CRISPR RNAs (crRNAs), that guide Cas proteins to the target DNA (Brouns et al. 2008). Marraffini and Sontheimer (2008) demonstrated that the target molecule is DNA (Marraffini and Sontheimer 2008). Subsequently, it was shown that CRISPR systems can function heterologously in other species (Sapranauskas et al. 2011). Subsequently, researchers from two groups successfully adapted CRISPR/Cas9 for genome editing in eukaryotic cells (Cong et al. 2013; Mali et al. 2013).

CRISPR/Cas9 nuclease has been developed that uses a specific guide RNA binding approach to recognize and cut a certain sequence (Gupta and Musunuru 2014). This approach has been found to be more effective than the previous approaches due to their high specificity and simplicity. CRISPR/Cas9 systems are dependent upon a combination of proteins and short RNAs to recognize the target DNA sequences, therefore, it doesn't need large DNA segments and thus can be designed easily (Mali et al. 2013). Myriad versions of the CRISPR/Cas9 can be generated to target a wide range of target sequences due to the involvement of short RNA sequences. CRISPR/Cas9 can also use multiple guide RNAs simultaneously, and thus, can target multiple sites at one time (Cong et al. 2013; Mali et al. 2013).

7.7 CRISPR/Cas9 Mediated Genome Editing in Plants

Application of CRISPR/Cas9 mediated genome editing in plants began in the year 2013. The first reported modification of a plant using a CRISPR/Cas system was by Shan et al. (2013). In the same year, Li et al. (2013) and Nekrasov et al. (2013) reported genome editing in *Arabidopsis* and *Nicotiana benthamiana* (Li et al. 2013; Nekrasov et al. 2013). A number of examples in recent years have demonstrated widespread application of CRISPR/Cas9 mediated genome editing for crop improvements.

7.7.1 Proof of Concept Studies

Some of the proof of concept studies on the successful use of CRISPR/Cas9 editing in various plants have been listed as follows: *Arabidopsis*, tobacco, sorghum and rice (Jiang et al. 2013; Li et al. 2013; Nekrasov et al. 2013), wheat (Jiang et al. 2013; Shan et al. 2013), rice (Barman et al. 2019; Jiang et al. 2013; Lee et al. 2019a, b, c; Li et al. 2016; Saika et al. 2019; Xu et al. 2015; Zhang et al. 2014; Zhou et al. 2014), tomato (Bari et al. 2019; Brooks et al. 2014; D'Ambrosio et al. 2018; Ito et al. 2015; Li et al. 2018a, b, c, d, 2019a, b; Ortigosa et al. 2019; Pan et al. 2016; Tashkandi et al. 2018; Veillet et al. 2019; Wang et al. 2019a, b; Yu et al. 2017), sweet orange (Jia and Wang 2014), maize (Char et al. 2017; Chen et al. 2018; Doll et al. 2019; Feng et al. 2016, 2018; Lee et al. 2019a, b, c; Li et al. 2017a, b; Liang et al. 2014; Malzahn et al. 2019; Qi et al. 2016; Shi et al. 2017; Svitashev et al. 2016; Zhu et al. 2016) and soybean (Al Amin et al. 2019; Bao et al. 2019a, b; Cai et al. 2015, 2018a, b, 2019; Chilcoat et al. 2017; Do et al. 2019; Du et al. 2016; Jacobs et al. 2015; Li et al. 2018a, b, c, d, 2019a, b; Liu et al. 2019; Michno et al. 2015; Sun et al. 2015). A more detailed list of recent publications in this field is provided in Tables 7.1 and 7.2.

Table 7.1 Application of CRISPR/Cas genome modifications in important crop plants

Host plant	Modification	Reference
Brassica napus (Oilseed rape)	CRISPR-Cas9 construct used to target two *ALCATRAZ* (*ALC*) homoeologs. Knocking out *ALC* increase shatter resistance to avoid seed loss during mechanical harvest	Braatz et al. (2017)
Brassica napus (Oilseed rape)	CRISPR/Cas9 mutagenesis of *BnCLV3* produced more leaves and multilocular siliques with a significantly higher number of seeds per silique and a higher seed weight than the wild-type and single mutant plants, potentially contributing to increased seed production	Yang et al. (2018)
Brassica napus (Oilseed rape)	Examined the mutation efficiency of the CRISPR/Cas9 method for 12 genes and also determined the pattern, specificity and heritability of these gene modifications in *B. napus*	Yang et al. (2017)
Oryza sativa (Rice)	Engineered CRISPR/Cas was active in creating DSBs when stably expressed in rice plants. CRISPR/Cas was used to achieve targeted genome modifications in both dicot and monocot plants	Feng et al. (2016)
Hordeum vulgare (Barley)	Employed SpCas9-mediated knockout strategy to functionally study *HvMORC1*, one of the Microrchidia (MORC) proteins that contribute in genome stabilization in monocotyledonous and dicotyledonous plants	Kumar et al. (2018)
Manihot esculenta (Cassava)	Used CRISPR/Cas9 technology in cassava, the *Phytoene desaturase* (*MePDS*) gene, which is a plant carotenoid biosynthetic enzyme was targeted in two cultivars	Odipio et al. (2017)
Manihot esculenta (Cassava)	CRISPR/Cas9-mediated genome editing was employed to generate *eif4e, ncbp-1, ncbp-2,* and *ncbp-1/ncbp-2* mutants in cassava cultivar 60,444 which delayed and attenuated cassava brown streak disease (CBSD) aerial symptoms and reduced severity and incidence of storage root necrosis as well	Gomez et al. (2019)
Gossypium hirsutum (Cotton)	CRISPR/Cas9 genome editing technology was used to knock out the cotton arginase gene (*GhARG*) to improve lateral root formation	Wang et al. (2018a, b)
Gossypium hirsutum (Cotton)	Cassette of sgRNA was designed to target Cotton leaf curl disease (CLCuD)-associated begomovirus complex and satellite molecules	Iqbal et al. (2016)
Cucumis sativus (Cucumber)	Generated a gynoecious cucumber line through CRISPR/Cas9-mediated mutagenesis of *CsWIP1*	Hu et al. (2017)

(continued)

Table 7.1 (continued)

Host plant	Modification	Reference
Cucumis sativus (Cucumber)	Development of virus resistance in cucumber using sgRNA technology by disrupting the function of the recessive *eIF4E* (eukaryotic translation initiation factor 4E) gene	Chandrasekaran et al. (2016)
Glycine max (soybean)	Demonstrated the effectiveness of CRISPR/Cas9 system in soybean by knocking-out a green fluorescent protein (GFP) transgene and modifying nine endogenous loci	Jacobs et al. (2015)
Glycine max (soybean)	Cas9-guide RNA (gRNA) was successfully applied to generate targeted mutagenesis, gene integration, and gene editing	Li et al. (2015)
Vitis vinifera (Grape)	Knockout of *VvWRKY52* in grape to increase the resistance to *Botrytis cinerea*	Wang et al. (2018a, b)
Nicotiana benthamiana	Resistance to tomato yellow leaf curl virus (TYLCV) using CRISPR/Cas9 system	Ali et al. (2015)
Brassica napus (Oilseed rape)	Modification of a fatty acid desaturase 2 gene (*FAD2*), which encodes an enzyme that catalyzes the desaturation of oleic acid, in *Brassica napus* cv. Westar using the CRISPR/Cas9 system	Okuzaki et al. (2018)
Citrus sinensis (Orange)	Demonstrated CRISPR/Cas9-mediated promoter editing strategy for generation of canker-resistant citrus cultivars	Peng et al. (2017)
Oryza sativa (Rice)	SgRNAs directed Cas9 to induce sequence-specific genome modifications in the rice	Shan et al. (2013)
Oryza sativa (Rice)	Created short life cycle in Kitaake, a *japonica* rice variety by targeting *CAO1* and *LAZY1* gene	Miao et al. (2013)
Oryza sativa (Rice)	Engineered guide RNAs were shown to direct the Cas9 nuclease for precise cleavage at the desired sites and introduce mutation (insertion or deletion) by error-prone non-homologous end-joining DNA repair	Xie and Yang (2013)
Oryza sativa (Rice)	Demonstrated multiplex genome editing and chromosomal-fragment deletion in stable transgenic rice plants	Xie et al. (2015)
Oryza sativa (Rice)	Designed CRISPR/Cas9 vector system, utilizing a plant codon-optimized Cas9 gene, for convenient and high-efficiency multiplex genome editing in rice	Xu et al. (2015)
Populus trichocarpa (Black cottonwood)	Four guide RNAs (gRNAs) were designed to target distinct poplar genomic sites of the phytoene desaturase gene 8 (*PtoPDS*) resulting in albino phenotypes	Fan et al. (2015)

(continued)

Table 7.1 (continued)

Host plant	Modification	Reference
Populus trichocarpa (Black cottonwood)	Two CoA ligase genes, *4CL1* and *4CL2*, associated with lignin and flavonoid biosynthesis, respectively were targeted using CRISPR/Cas9 editing	Zhou et al. (2015a, b)
Oryza sativa (Rice)	CRISPR/Cas9 to generate targeted double-strand breaks and to deliver an RNA repair template for homology directed repair	Butt et al. (2017)
Oryza sativa (Rice)	RNAi expression element is incorporated into a CRISPR/Cas9 vector, the activity and presence of the T-DNA in the transgenic plants monitored based on RNAi	Lu et al. (2017)
Oryza sativa (Rice)	Disrupted the fatty acid desaturase 2(*OsFAD2-1*) gene which catalyzes the conversion of oleic acid to linoleic acid in rice by CRISPR/Cas9-mediated targeted mutagenesis resulting in production of rice with improved fatty acid composition	Abe et al. (2018)
Oryza sativa (Rice)	Inactivation of the Cs^+ -permeable K^+ transporter *OsHAK1* with the CRISPR-Cas system dramatically reduced Cs^+ uptake by rice plants	Nieves-Cordones et al. (2017)
Oryza sativa (Rice)	Used CRISPR/Cas9-mediated multiplex to develop early maturing varieties of rice by targeting *Hd2*, *Hd3* and *Hd5* involved in flowering suppression	Li et al. (2017a, b)
Oryza sativa (Rice)	Development of new *indica* rice lines with low Cd accumulation and no transgenes by knocking out the metal transporter gene *OsNramp5* using CRISPR/Cas9 system	Tang et al. (2017)
Oryza sativa (Rice)	Replaced the japonica *NRT1.1B* allele with the indica allele, in just one generation, using CRISPR/Cas9 gene-editing technology	Li et al. (2018a, b, c, d)
Oryza sativa (Rice)	Utilized CRISPR/Cas9 technology to edit pyrabactin resistance 1 (*PYR1*)/PYR1-like (*PYL*) genes to promote rice growth and productivity	Miao et al. (2018)
Oryza sativa (Rice)	CRISPR/Cas9-based genome editing was used to knockout the *SaF* and *SaM* alleles of an *indica* rice line to create hybrid-compatible lines	Xie et al. (2017)
Oryza sativa (Rice)	Created high-amylose rice through CRISPR/Cas9-mediated editing of starch branching enzyme IIb (*SBEIIb*)	Sun et al. (2017)

(continued)

Table 7.1 (continued)

Host plant	Modification	Reference
Oryza sativa L (Rice)	Knocking out one or two of the three indica rice allele *Sc-i* copies by CRISPR/Cas9 rescues *japonica* rice allele *Sc-j* expression and male fertility	Shen et al. (2017)
Oryza sativa (Rice)	Used CRISPR/Cas9 to introduce a loss-of-function mutation into the *Waxy* gene in two widely cultivated elite japonica varieties	Zhang et al. (2018a, b)
Oryza sativa (Rice)	Mutant GFP gene contained target sites in its 5′ coding regions were successfully cleaved by a CAS9/sgRNA complex that, along with error-prone DNA repair, resulted in creation of functional GFP genes	Jiang et al. (2013)
Oryza sativa (Rice)	Improvement of rice blast resistance by engineering a CRISPR/Cas9 SSN (C-ERF922) targeting the *OsERF922* gene in rice	Wang et al. (2016)
Solanum lycopersicum (Tomato)	*Agrobacterium tumefaciens*-mediated CRISPR/Cas9 system transformation method was used for obtaining tomato *ALC* gene mutagenesis and replacement, in absence and presence of the homologous repair template resulting in breeding tomatoes with long shelf life	Yu et al. (2017)
Solanum lycopersicum (Tomato)	Created targeted mutations in phytoene desaturase (*pds*), an important gene in the carotenoid biosynthesis pathway	Parkhi et al. (2018)
Solanum lycopersicum (Tomato)	Targeted point mutagenesis at genomic regions specified by single-guide RNAs (sgRNAs)	Shimatani et al. (2017)
Solanum lycopersicum (Tomato)	To increase GABA content in tomato by deleting the autoinhibitory domain of *SlGAD2* and *SlGAD3* using CRISPR/Cas9	Nonaka et al. (2017)
Solanum lycopersicum (Tomato)	Used CRISPR/Cas9 system to introduce somatic mutations effectively into *SlIAA9*, gene controlling parthenocarpy	Ueta et al. (2017)
Solanum lycopersicum (Tomato)	Developed resistance to the powdery mildew fungal pathogen using the CRISPR/Cas9 technology	Nekrasov et al. (2017)
Solanum lycopersicum (Tomato)	Used CRISPR/Cas9 to target mutations to the *PROCERA* encoded *DELLA* domain of tomato to create gibberellins responsive dominant dwarf	Tomlinson et al. (2019)
Solanum lycopersicum (Tomato)	Promoter was inserted upstream of a gene controlling anthocyanin biosynthesis, resulting in overexpression and ectopic accumulation of pigments in tomato tissues	Čermák et al. (2015)

(continued)

Table 7.1 (continued)

Host plant	Modification	Reference
Citrus sinensis (Sweet orange)	Targeted genome modification in citrus using the Cas9/sgRNA system to the study of citrus gene function and for targeted genetic modification	Jia et al. (2017)
Panicum virgatum (Switchgrass)	Developed CRISPR/Cas9 genome editing system to target a key enzyme involved in the early steps of monolignol biosynthesis to create switchgrass knock-out mutant plants with decreased lignin content and reduced recalcitrance	Park et al. (2017)
Theobroma cacao (Cacao tree)	Application of genome editing technology in cacao, using Agrobacterium-mediated transient transformation to introduce CRISPR/Cas9 components into cacao leaves and cotyledon cells	Fister et al. (2018)
Triticum aestivum (Wheat)	Use of sgRNAs for sequence-specific CRISPR/Cas-mediated mutagenesis and gene targeting in wheat	Shan et al. (2014)
Triticum aestivum (Wheat)	Knock-down of *TaEDR1* by virus-induced gene silencing or RNA interference enhanced resistance to powdery mildew, indicating that *TaEDR1* negatively regulates powdery mildew resistance in wheat	Zhang et al. (2017)
Triticum aestivum (Wheat)	Designed two sgRNAs to target a conserved region adjacent to the coding sequence for the 33-mer in the α-gliadin genes which produced low-gluten, transgene-free wheat lines which could suppress Coeliac disease	Sánchez-León et al. (2018)
Triticum aestivum (Wheat)	Used CRISPR/Cas9 in hexaploid bread wheat to introduce targeted mutations in the three homoeoalleles that encode MILDEW-RESISTANCE LOCUS (MLO) proteins	Wang et al. (2014)
Triticum aestivum (Wheat)	Applied CRISPR/Cas9 genome editing system in wheat protoplast to conduct the targeted editing of stress-responsive transcription factor genes	Kim et al. (2018)
Zea mays (Maize)	Demonstrated targeted mutagenesis in *Zea mays* using TALENs and the CRISPR/Cas system induced targeted mutations in *Z. mays* protoplasts resulting in genome modification in maize	Liang et al. (2014)

7.7.2 Enhancement of Yield

Increasing the yield is a prime focus for crop improvement considering the aim of global food security. Few studies demonstrating the use of CRISPR/Cas9 for improvement of yield is described here. Flowering is governed by season and day-length in many plants and this is a prime factor that determines the suitability of a crop plant at a particular geographic location. Modification of the photoperiod

Table 7.2 CRISPR/Cas9 editing in a model plant *Arabidopsis thaliana*

Modification	Reference
Examined several plant generations with seven genes at 12 different target sites to determine the patterns, efficiency, specificity, and heritability of CRISPR/Cas-induced gene mutations or corrections in *Arabidopsis thaliana*	Feng et al. (2014)
Nuclease-induced targeted mutagenesis events in the *ADH1* and *TT4* genes of *Arabidopsis thaliana* resulting in stable inheritance of nuclease-induced targeted mutagenesis events	Fauser et al. (2014)
Demonstrating the efficiency of Cas9/sgRNA in causing modification of a chromosomally integrated target reporter gene	Jiang et al. (2014)
Co-expressed the plant codon-optimized *SpCas9* (*pcoCas9*) and a gRNA targeting the *Arabidopsis thaliana PDS3* (*PHYTOENE DESATURASE*) gene, to disrupt *PDS3* to abolish carotenoid biosynthesis and promote chlorophyll oxidation leading to a photobleached phenotype	Li et al. (2013)
Demonstrated that the Cas9 nuclease is able to induce heritable mutations in *Arabidopsis thaliana*	Schiml et al. (2014)
Site-directed mutagenesis of the Arabidopsis nuclear genome that efficiently generates heritable mutations using the RNA-guided endonuclease (RGEN) derived from bacterial CRISPR/Cas9 system	Hyun et al. (2015)
CRISPR/Cas9 technology to introduce sequence-specific deleterious point mutations at the *eIF(iso)4E* locus in *Arabidopsis thaliana* to successfully engineer complete resistance to Turnip mosaic virus (TuMV)	Pyott et al. (2016)
Agrobacterium tumefaciens was used for delivery of genes encoding Cas9, sgRNA and a non-functional, mutant green fluorescence protein (GFP) to *Arabidopsis thaliana*	Jiang et al. (2013)
Engineered CRISPR/Cas was active in creating DSBs when transiently expressed in *Arabidopsis* protoplasts and stably expressed in transgenic *Arabidopsis thaliana*	Feng et al. (2013)

response can be highly beneficial considering its potential to break the geographic specificity barrier. In a study reported by Soyk et al., it was demonstrated that the loss of day-length-sensitive flowering in tomato is driven by the florigen paralog and flowering repressor *SP5G* (Soyk et al. 2017). CRISPR/Cas9-engineered mutations in *SP5G* caused faster flowering resulting in early yield and demonstrated the power of gene editing.

The use of multiple guide RNAs for complex genome editing is a challenging objective. Cai et al. (2018a, b) demonstrated the use of CRISPR/Cas9 technology for targeted deletions of DNA fragments in *GmFT2a* and *GmFT5a* in soybean. Furthermore, they have demonstrated the heritability of such changes in the T_2 generation with late-flowering phenotype (Cai et al. 2018a, b).

Rice is an important staple food in various parts of the world. Hybrid rice breeding is considered an important strategy for rice improvement. Production of a male-sterile line is highly advantageous for rice breeding. The photo- and temperature-sensitive male sterile line is required for two-line hybrid rice breeding (Barman et al. 2019). In a study, Zhou et al. (2016)demonstrated an efficient rice thermo-sensitive genic male sterility (TGMS) cultivating system. CRISPR/Cas9 was utilized to knock out *TMS5* and develop 11 lines within a short span of one year (Zhou et al. 2016).

In another study, grain yield was enhanced by modifying multiple sets of regulatory genes by CRISPR/Cas9 (Li et al. 2016). From previous studies, it was known that the *Gn1a*, *DEP1*, *GS3* and *IPA1* controls grain yield through enhancing the grain number, panicle architecture, grain size and plant architecture, respectively. The plants mutated in the above four genes by CRISPR/Cas9 showed an enhancement of grain yield similar to those described in some previous reports. The T_2 generation of the *gn1a*, *dep1*, and *gs3* mutants demonstrated increased grain number, dense and erect panicles, and larger grain size, respectively (Li et al. 2016).

The use of CRIPSR/Cas9 technology for improving crop yield has also been reported in tomato by the targeted editing of the *cis*-regulatory element affecting fruit size, inflorescence architecture and growth habit (Rodríguez-Leal et al. 2017). An increase of up to 70% in the size of tomato fruits was reported in this communication. This methodology therefore provides a foundation for dissecting the interactions between alterations in gene regulations and their effects on quantitative traits towards the ultimate goal of yield improvement.

7.7.3 *Quality Enhancement*

Ma et al. (2015) utilized multiple sgRNA cassettes for multiplex genome editing, resulting in the multiple, simultaneous changes in the *OsWaxy* gene, causing a five-fold reduction of amylose content in the resultant rice plants. Knocking out the *Waxy* gene towards achieving the same goal has also been reported recently by Zhang et al. (2018a, b). For improving the nutritional quality of wheat, low-gluten lines were developed by multiplex knockout of α-gliadin genes (Sánchez-León et al. 2018). This technology has also been used for the production of nutritionally desirable high-amylose containing rice by editing *SBEI* and *SBEIIb* through CRISPR/Cas9 technology (Sun et al. 2017). As well, a reduction of the anti-nutritional compound Phytic acid was achieved through the targeting of inositol phosphate kinase (*ZmIPK*) gene, a key enzyme in their biosynthetic pathway (Liang et al. 2014). Lawrenson et al. (2015) demonstrated the utility of CRISPR/Cas9 system in targeting the important agronomic trait of dormancy, by editing the ABA-inducible plasma membrane protein gene, *HvPM19* (Lawrenson et al. 2015).

Tomato plants producing increased levels of γ-aminobutyric acid (GABA, a non-proteinogenic amino acid with hypotensive properties) was generated employing NHEJ (Nonaka et al. 2017). The production of tomato fruits of different colors (purple, yellow, pink) catering to differing consumer preferences, through targeting *Anthocyanin 2*, *phytoene synthase 1* or *MYB transcription factor12* genes have also been produced through the application of CRISPR/Cas9 technology (Čermák et al. 2015; Filler Hayut et al. 2017). The amount of the bioactive compound lycopene in tomato fruits (thought to lower the incidence of cancer and cardiovascular diseases), has been increased over fivefold employing CRISPR/Cas9 technology to target carotenoid biosynthesis pathway (Li et al. 2018a, b, c, d). On the other hand, Tang et al. (2017) knocked out the metal transporter gene (*OsNramp5*) resulting in a dramatic decrease of the toxic heavy metal Cadmium concentration in the grains of *indica* rice without compromising yield (Tang et al. 2017).

7.7.4 Biotic Stresses/Disease Resistance

Targeted mutagenesis of CRISPR/Cas9 mediated deletion of several nucleotides from the promoter region of *OsSWEET11* and *OsSWEET14* modulating resistance to bacterial leaf blight was reported by Jiang et al. (2013). Improved resistance against bacterial blight has been achieved in rice by modifying *OsSWEET13* sugar transporter gene (Zhou et al. 2015a, b), targets of the pathogens' transcription activator-like affectors (TALEs), which are required for their virulence. As well, Wang et al. (2016) knocked out the ERF transcription factor gene (*OsERF922*) imparting enhanced resistance against the fungal pathogen responsible for rice blast disease (Wang et al. 2016). Recently, CRISPR/Cas9 technology has also been used to generate canker resistant citrus varieties by editing the *CsLOB1* gene (Jia et al. 2017). In bread wheat, simultaneous editing of three homeoalleles of the *MLO* gene conferred resistance to powdery mildew (Wang et al. 2014). Similarly, Zhang et al. (2017) reported simultaneous editing of three *TaEDR1* homeologs, enhancing resistance to the same pathogen (Zhang et al. 2017). As well, transgene-free tomato with enhanced powdery mildew resistance has been achieved using CRISPR/Cas9 mediated editing of *SlMlo1* gene (Nekrasov et al. 2017), and Kumar et al. recently (2018) reported the generation of gene-edited barley genotypes (SpCas9-edited hvmorc1-KO barley with mutated microrchidia proteins) imparting increased resistance to fungal pathogens (Kumar et al. 2018).

The application of the CRISPR/Cas9 system for gene editing towards modulating virus resistance has been reported in a number of cases (Ali et al. 2015; Baltes et al. 2015; Iqbal et al. 2016; Ji et al. 2015). Ali et al. (2015) used this system to generate *Nicotiana bentahamiana* plants resistant against Beet curly top virus, Tomato yellow leaf curl virus and Merremia mosaic virus, while Bean yellow dwarf virus (Baltes et al. 2015), and BSCTV (Ji et al. 2015) has also been targeted. Broad-spectrum Geminivirus resistance has also been demonstrated against several begomoviruses (Zaidi et al. 2016). Resistance to begomovirus in cotton (Iqbal et al. 2016) and the production of cucumber genotypes resistant to cucumber vein yellowing virus, Zucchini yellow mosaic virus and Papaya ringspot mosaic virus cucumber has also been reported (Chandrasekaran et al. 2016). However, a note of caution needs to be recognized since a recent study (Mehta et al. 2019) aiming to engineer Geminivirus resistance in cassava plants, observed the enhanced evolution of mutated virus, resistant to CRISPR/Cas9 cleavage. Therefore, a careful assessment of this technology in relation to biosafety risks needs to be undertaken to circumvent the inadvertent effects of this technology.

7.7.5 Abiotic Stresses

Herbicide (bentazon and sulfonylurea) resistant rice plants were generated through *Agrobacterium*-mediated transformation and CRISPR/Cas9 technology (Xu et al. 2014). Similarly, Sun et al. (2016) generated herbicide-tolerant rice plants by mutating the branched-chain amino acid biosynthesis gene (acetolactate synthase, *ALS*) employing CRISPR/Cas9 technology (Sun et al. 2016). On the other hand, Li et al. (2016) produced glyphosate-resistant rice plants through targeting *OsESPS* gene using this technology (Li et al. 2016). As well, Tian et al. (2018) recently reported the production of herbicide-resistant watermelon varieties through the targeted mutation of the *ALS* gene employing CRISPR-Cas9 technology (Tian et al. 2018). Recently, editing the low-temperature sensitivity regulating transcription factor *TIFY1* has been demonstrated (Huang et al. 2017), which can shed light on the regulation of low-temperature sensitivity and adaptation.

7.7.6 Drought Tolerance

Drought-tolerant maize variants have been generated by editing the weak, endogenous promoter of *AGROS8*, the negative regulator of ethylene response, using CRISPR/Cas9 technology (Shi et al. 2017). On the other hand, Wang et al. (2017) reported the MAPKs edited tomato plants being more sensitive to drought when compared to the non-edited controls (Wang et al. 2017). However, they screened just two transgenic line, and it is entirely possible that that the production and screening of additional lines could lead to the identification of genotypes with enhanced resistance to drought, since most of the mutations/changes produced, by nature, are detrimental. Similarly, Li et al. (2019a, b) have also reported the generation of tomato plants with reduced tolerance to drought due CRISPR/Cas9 mediated mutagenesis of the *NPR1* (non-expressor of pathogenesis-related gene 1) (Li et al. 2019a, b). However, *NPR1* is a master regulator of plant defense response, but not much information is available on its role in modulating abiotic stresses, although another pathogenesis-related protein (PR10) is known to be involved in imparting drought- and salt-tolerance (Srivastava et al. 2006). Several recent publications (Chen et al. 2019; Paixão et al. 2019) have reported the enhancement of drought and salt tolerance employing CRISPR/Cas9 technology. Paixão et al. (2019) fused the dCAS9 with the gene activator histone acetyltransferase (HAT) targeting the Abscisic acid-responsive element binding protein 1 (AREB1), a key regulator of drought stress response in plants (Paixão et al. 2019). The resultant *Arabidopsis* plants exhibited improved survival rate after the imposition of drought stress. In another such example, Chen et al. (2019) established the editing of transcription repressors DPA4 and SOD7 enhancing abiotic salt and drought tolerance along with an increase in seed size through the regulation of inflorescence architecture in *Arabidopsis*.

In a recent study, Bhowmik et al. (2018) incorporated the use of CRSPR-Cas9 technology with wheat microspore culture to demonstrate the possibility of exploiting haploid breeding technology and targeted gene editing (Bhowmik et al. 2018). They were able to transform both endogenous and exogenous (reporter) genes and regenerate microspore-derived plants using this technology. Similarly, Hensel et al. (2018) reported the regeneration of haploids from pollen grains of primary transformants, demonstrating the combined utility of haploid technology and gene editing (Hensel et al. 2018). Lu and Zhu (2017) achieved enhanced nitrogen use efficiency by targeting *NTR1, 1B* genes through targeted base editing, while Zong et al. (2017) modulated senescence and cell death by targeting *OsCDC48* employing the same technology (Lu and Zhu 2017; Zong et al. 2017). The shelf-life of tomato fruits has also been increased through the targeted mutagenesis and gene replacement of the *ALC* gene in tomato (Yu et al. 2017).

7.8 Intellectual Property Rights and Commercialization Potential

Genome editing has many advantages over conventional transgenic techniques for crop improvement. Unlike transgenic techniques that rely on the introduction of a foreign genetic material to induce variations, genome-editing depends on the host's own repair system for inducing such changes. Also, these changes are quite similar to the naturally occurring mutations *in planta*. Genomic editing, on the other hand, is the relatively simpler approach. A lot of nucleases or molecular scissors can be produced and edited as per requirements using latest technologies. The most critical part, however, is to identify the right location to induce those changes. Genome editing exploits plants' innate features and make them more attractive and may have a more favorable view of the general population towards their acceptance since they are not incorporating heterologous genes from a foreign individual, a common feature of traditional GM plants.

The number of patent applications related to genome-editing techniques has been increased by 15 folds since 2005 (Brinegar et al. 2017). A list of recent patents in this field is listed in Table 7.3. Interestingly, most of these IP applications came from academic institutions rather than industrial giants, which indicate the advent of academic initiatives and entrepreneurship in the field of genomics and advanced biotechnology. Nevertheless, most of the patent market is captured by the United States due to its relatively flexible regulatory system. Europe has been left behind in the commercialization of genome-edited products due to its more rigid regulatory system, in which scientists are allowed to conduct research, but not permitted to commercialize favorable outcomes. In the last few years alone, US has invested over $1 billion in gene-editing start-up companies (Brinegar et al. 2017).

Table 7.3 Patents related to CRISR-CAS genome editing and its applications

Patent description	Inventors	Patent no.
CRISPR-CAS systems and methods for altering expression of gene products, structural information and inducible modular CAS enzymes	Feng Zhang, Bernd Zetsche	10,377,998
Compositions and methods for the expression of CRISPR guide RNAs using the H1 promoter	Vinod Jaskula-Ranga, Donald Zack	9,907,863, 10,369,234
CRISPR enabled multiplexed genome engineering	Ryan T. Gill, Andrew Garst	9,982,278, 10,240,167, 10,266,849, 10,351,877, 10,364,442
Methods and systems for identifying CRISPR/Cas off-target sites	Thomas James Cradick, Gang Bao, Peng Qiu	10,354,746
CRISPR-based methods and products for increasing frataxin levels and uses thereof	Jacques P. Tremblay, Pierre Chapdelaine, Joel Rousseau	10,323,073
Reduction of amyloid beta peptide production via modification of the *APP* gene using the CRISPR/Cas system	Jacques P. Tremblay, Joel Rousseau, Pierre Chapdelaine	10,280,419
CRISPR effector system based diagnostics	Omar Abudayyeh, James Joseph Collins, Jonathan Gootenberg, Feng Zhang, Eric S. Lander	10,266,886 10,266,887
CRISPR/Cas-mediated genome editing to treat EGFR-mutant lung cancer	Huibin Tang, Joseph B. Shrager	10,240,145
Gene editing in the oocyte by Cas9 nucleases	Ralf Kuhn, Wolfgang Wurst, Oskar Ortiz Sanchez	9,783,780 10,214,723
Therapeutic uses of genome editing with CRISPR/Cas systems	Kiran Musunuru, Chad A. Cowan, Derrick J. Rossi	10,208,319
Compositions and methods directed to CRISPR/Cas genomic engineering systems	Fangting Wu	10,202,619
Engineered CRISPR-Cas9 nucleases with altered PAM specificity	J. Keith Joung, Benjamin Kleinstiver	9,944,912, 10,202,589
CRISPR-related methods and compositions with governing gRNAS	Feng Zhang, Deborah Palestrant, Beverly Davidson, Jordi Mata-Fink, Edgardo Rodriguez, Alexis Borisy	9,834,791, 10,190,137
Methods for screening bacteria, archaea, algae, and yeast using CRISPR nucleic acids	Rodolphe Barrangou, Kurt M. Selle	10,136,649
Engineered CRISPR-Cas9 nucleases	J. Keith Joung, Benjamin Kleinstiver, Vikram Pattanayak	9,512,446 9,926,545 9,926,546 10,093,910

(continued)

Table 7.3 (continued)

Patent description	Inventors	Patent no.
Methods for increasing CAS9-mediated engineering efficiency	Peter Sean Cameron, Rachel E. Haurwitz, Andrew P. May, Christopher H. Nye, Meganvan Overbeek	9,970,030
Compositions and methods of engineered CRISPR-CAS9 systems using split-nexus CAS9-associated polynucleotides	Paul Daniel Donohoue, Andrew Paul May	9,580,727 9,745,600 9,970,026 9,970,027
CRISPR/CAS-related methods and compositions for treating leber's congenital amaurosis 10 (LCA10)	Morgan L. Maeder, David A. Bumcrot, Shen Shen	9,938,521
CRISPR oligonucleotides and gene editing	Namritha Ravinder, Korbinian Heil, Yizhu Guo, Xiquan Liang, Robert Potter, Sanjay Kumar	9,879,283
Method for fragmenting genomic DNA using CAS9	Gusti Zeiner, Derek Lee Lindstrom, Brian Jon Peter, Robert A. Ach	9,873,907
CRISPR hybrid DNA/RNA polynucleotides and methods of use	Andrew Paul May, Paul Daniel Donohoue	9,771,601 9,580,701 9,650,617 9,688,972 9,868,962
Methods for engineering T cells for immunotherapy by using RNA-guided CAS nuclease system	Philippe Duchateau, Andre Choulika, Laurent Poirot	9,855,297
CAS9-based isothermal method of detection of specific DNA sequence	Carsten-Peter Carstens	9,850,525
CRISPR-Cas component systems, methods and compositions for sequence manipulation	Feng Zhang	8,795,965 9,840,713
CRISPR-based compositions and methods of use	Michael Allen Collingwood, Ashley Mae Jacobi, Garrett Richard Rettig, Mollie Sue Schuber, Mark Aaron Behlke	9,840,702
CRISPR-Cas component systems, methods and compositions for sequence manipulation	Feng Zhang, David Benjamin Turitz, Luciano Marraffini, David Bikard, Wonsan Jiang, Neville Espi Sanjana	9,822,372
Method of making a deletion in a target sequence in isolated primary cells using Cas9 and two guide RNAs	Kiran Musunuru, Chad A. Cowan, Derrick J. Rossi	9,822,370
Engineered nucleic acid-targeting nucleic acids	Paul Daniel Donohoue, Andrew Paul May	9,677,090 9,816,081 9,816,093
Compositions and methods of nucleic acid-targeting nucleic acids	Andrew Paul May, Rachel E. Haurwitz, Jennifer A. Doudna, James M. Berger, Matthew Merrill Carter, Paul Daniel Donohoue	9,725,714 9,803,194 9,809,814

(continued)

Table 7.3 (continued)

Patent description	Inventors	Patent no.
CRISPR enzymes and systems	Feng Zhang, Bernd Zetsche, Ian Slaymaker, Jonathan Gootenberg, Omar O. Abudayyeh	9,790,490
Methods of using engineered nucleic-acid targeting nucleic acids	Paul Daniel Donohoue, Andrew Paul May	9,745,562
CRISPR/Cas systems for genomic modification and gene modulation	Fangting Wu	9,738,908
Use of cationic lipids to deliver CAS9	David R. Liu, John Anthony Zuris, David B. Thompson	9,737,604
Compositions and methods for synthetic gene assembly	Esteban Toro, Sebastian Treusch, SiyuanChen, Cheng-Hsien Wu	9,677,067
Methods and compositions for producing double allele knock outs	Bo Yu, James Larrick	9,663,782
Methods and compositions for regulation of zinc finger protein expression	Steven Froelich, Philip D. Gregory, H. Steve Zhang	9,624,498
Nuclease-mediated DNA assembly	Chris Schoenherr, John McWhirter, Corey Momont, Caitlin Montagna, Lynn Macdonald, Gregg S. Warshaw, Jose F. Rojas, Ka-Man Venus Lai, David M. Valenzuela, Andrew J. Murphy	9,580,715
Using truncated guide RNAs (tru-gRNAs) to increase specificity for RNA-guided genome editing	J. Keith Joung, Jeffry D. Sander, Morgan Maeder, Yanfang Fu	9,567,604
Using RNA-guided FokI nucleases (RFNs) to increase specificity for RNA-guided genome editing	J. Keith Joung, Shengdar Tsai	9,567,603
Cas9-recombinase fusion proteins and uses thereof	David R. Liu, John Paul Guilinger, David B. Thompson	9,388,430
Cas9-FokI fusion proteins and uses thereof	David R. Liu, John Paul Guilinger, David B. Thompson	9,322,037
Evaluation and improvement of nuclease cleavage specificity	David R. Liu, John Paul Guilinger, Vikram Pattanayak	9,322,006
Compositions and methods of nucleic acid-targeting nucleic acids	Andrew Paul May, Rachel E. Haurwitz, Jennifer A. Doudna, James Berger Matthew Merrill M. Carter, Paul Donohoue	9,260,752
Methods and compositions for the targeted modification of a genome	David Frendewey, Wojtek Auerbach, Ka-Man Venus Lai, David M. Valenzuela, George D. Yancopoulos	9,228,208
Switchable gRNAs comprising aptamers	David R. Liu, Johnny Hao Hu	9,228,207

(continued)

Table 7.3 (continued)

Patent description	Inventors	Patent no.
Methods for identifying a target site of a Cas9 nuclease	David R. Liu, Vikram Pattanayak	9,163,284
CRISPR-Cas systems and methods for altering expression of gene products	Feng Zhang	8,945,839
CRISPR-Casnickase systems, methods and compositions for sequence manipulation in eukaryotes	Le Cong, Feng Zhang	8,889,356 8,932,814
Engineering of systems, methods and optimized guide compositions for sequence manipulation	Feng Zhang, Le Cong, Patrick Hsu, Fei Ran	8,906,616
Engineering and optimization of improved systems, methods and enzyme compositions for sequence manipulation	Feng Zhang, Fei Ran	8,865,406 8,889,418 8,895,308
CRISPR-Cas component systems, methods and compositions for sequence manipulation	Le Cong, Feng Zhang	8,871,445
CRISPR-Cas component systems, methods and compositions for sequence manipulation	Feng Zhang	8,795,965
CRISPR-Cas systems and methods for altering expression of gene products	Feng Zhang	8,697,359 8,771,945

7.9 Regulatory Consideration for Genome-Edited Plants

Emerging technologies have always been represented to be controversial issues with the public, especially when it comes to the applications of biotechnology, where it is considered as tampering with nature. The commercial application of DNA technologies in 1994 with the development of the FLAVR SAVR™ tomato caused mixed public reactions that varied between curiosities and concerns (Bruening and Lyons 2000). By 2015, only 37% of the public considered genetically engineered (GE) foods are safe, compared to 88% of scientists from a wide range of disciplines (Funk and Rainie 2015).

The products of biotechnology go under a lengthy and thorough vetting process to prove their safety. In the US, the regulations mostly focus on the product itself rather than the development process. It is largely driven by public perceptions and commitments (Wolt and Wolf, 2018) such as the case of deregulation of glyphosate-resistant alfalfa.

The emergence of gene editing has challenged the regulators with the fact that genome editing is a process of modifying crops or organisms that do not contain foreign or recombinant DNA as the GM plants (Wolt and Wolf 2018). The technology brought public uncertainties and raised discussions in many countries regarding their regulation and their applications (O'Keefe et al. 2015). While the existing regulations do not encumber the commercialization of genome-edited crops, it has certainly brought into focus the need to regulate, due to its high innovative potentials and future products (Huang et al. 2016; Jones 2015; National Academies of Sciences and Medicine 2016; Wolt et al. 2016; Wolt and Wolf 2018). The regulatory frameworks take into account the development process or the characteristics of the product developed as well as the differences and similarities of genome-edited and GMOs. The discussion addresses the regulatory framework that includes monitoring, traceability, labelling and socio-economic issues (Eckerstorfer et al. 2019).

United States: In March 2018, the US Department of Agriculture excluded genome-edited plants from regulatory oversight altogether. By contrast, the Court of Justice of the European Union ruled in July 2018 that it would treat gene-edited crops as genetically modified organisms, subject to stringent regulation. The United States Department of Agriculture Animal and Plant Health Inspection Service (USDA-APHIS) has approved a CRISPR-edited non-browning mushroom in 2016. This was followed by the approval of the CRISPR-edited corn, soybeans, tomatoes, pennycress, and *Camelina*. These crops are not yet in the market and may be regulated by the Food and Drug Administration (FDA) and the Environmental Protection Agency (EPA) at a later date. GMOs from the older modification techniques have been very costly for the companies to be marketed, due to the tight regulations worldwide and the length of time to be approved (Taylor 2019). The approvals may cost up to $130 million per genetic change and can take up to a decade to complete the process of commercialization. That leaves the companies to consider only crops that are profitable such as corn or soybeans and only big companies dominate the market of modified crops. The advent of genome editing technology has opened the market for smaller companies. In 2016, lawmakers passed The US National Bioengineered Food Disclosure Law, which requires manufacturers to label products containing "bioengineered" food that contains genetic material that has been modified through in vitro recombinant DNA techniques; and for which the modification could not otherwise be obtained through conventional breeding or found in nature. In 2018, the US Secretary of Agriculture stated that the USDA would not regulate crops that "could otherwise have been developed through traditional breeding techniques as long as they are not plant pests or developed using plant pests," such as CRISPR-edited crops (Taylor 2019).

Europe: The European Court of Justice (ECJ) judgment of 25 July 2018, C-528/16, stated that the genome-edited organisms are considered as genetically modified organisms and are subject to the same regulations. This regulation is based on the uncertainties that the genome-edited organisms cannot be distinguished from the conventionally bred organisms and therefore they can be under the same law of GMO or exempted from it. The European framework is based on risk assessment traceability and labeling. The ECJ argued that the products made by gene-editing tools such as CRISPR are fundamentally different from transgenic techniques, therefore should not be included under the GMO umbrella unless they contain transgenic material. On the other hand, some researchers and environmentalists groups argued that gene-editing techniques carry their own risks, and should therefore be fully tested and assessed for their safety as genetically engineered organism. In this regards, the European Commission's chief scientific advisors stated that it is impossible to determine whether a mutation occurred through gene editing or through natural mutation, therefore it is advisable to focus on the safety of the product, and not on the process by which it was created. The uncertainties imposed limitations on research and development mainly on plant biotechnology corporate companies (Smyth and Lassoued 2019; Zimny et al. 2019) and lead to loss of competitiveness for Europe's agricultural industry with high economic consequences (Kalaitzandonakes et al. 2014; Ryan and Smyth 2012). The EU Commission and the European agencies are to address update for the existing law that may include providing unique identifier for genome-edited organisms (Wasmer 2019).

Australia: In Australia, research using CRISPR/Cas9 technologies was restricted and considered the same as conventional genetics modifications, therefore follows the same rules. The approval of the modifications is required by the Office of Gene Technology Regulator (OGTR) for its biosafety. In an update of the regulations, the government decided not to regulate the use of gene-editing techniques in plants, animals and human cell lines that do not introduce new genetic material. The updated regulations allow scientists to use some genome-editing techniques in plants and animals without government approval. The Australian regulator says that genetic edits made without templates are no different from changes that occur in nature, and therefore do not pose an additional risk to the environment and human health. Gene-editing technologies that do use a template, or that insert other genetic material into the cell, will continue to be regulated by the OGTR.

7.10 Public Perception

The general public has so many misconceptions regarding genome-edited crops. Most of their perceptions are derived from media and false information. They are highly skeptical about the quality and safety of genome-edited crops. Approximately 37% of public believe that GM food is safe to eat, rest are either unsure or have a negative perception. Although, all of these fears are not baseless, however, there is a need to make people more aware about the genetically engineered techniques, their advantages, as well as their possible disadvantages, so that people can get real knowledge about what they are consuming and how it can affect their health. Public awareness regarding genetically engineered crops is a must for the sustained development of future advances in agriculture biotechnology. Scientists therefore should be more engaging the public by informing them of the scientific basis of their achievements, how those changes are brought about. This engagement of the scientific community with the public at large, in a more lucid way, should help clear the general misconception regarding the safety (as well as the urgent requirement in lieu of global food security) concerning the genetically modified foods.

7.11 Conclusion and Perspectives

Genome editing is a more progressive and precise technique to bring desirable changes in crop plants. Technological advancements are necessary towards producing crops with higher quality and better traits. In comparison to plasmid-mediated stable transformants, direct delivery of purified Cas9 protein with guide RNA into plant cells shows better efficiency and reduced off-target effects (Kanchiswamy et al. 2015). Following regeneration from edited cells, the genome editing complex is degraded in the recipient cells. This phenomenon may be useful in overcoming regulatory hurdles since the resultant genome-edited plants do not contain any transgenic sequences. The biggest hurdle in plant biotechnology is to overcome the regulatory and ethical hurdles sabotaging the innovation and development of new plant breeding techniques. The US and Europe are two of the biggest hubs for research and development, and they have to find a common ground to facilitate crop improvement. It is also crucial to switch from process-based regulatory system to product-based analysis. The end product and its safety is more important than the process used to develop such products. Therefore, the future success of genome editing will largely depend upon their acceptance from regulatory bodies and the general public.

References

Abe K, Araki E, Suzuki Y, Toki S, Saika H (2018) Production of high oleic/low linoleic rice by genome editing. Plant Physiol Biochem 131:58–62

Al Amin N, Ahmad N, Wu N, Pu X, Ma T, Du Y, Bo X, Wang N, Sharif R, Wang P (2019) CRISPR-Cas9 mediated targeted disruption of *FAD2–2* microsomal omega-6 desaturase in soybean (*Glycine max* L). BMC Biotechnol 19:9

Ali Z, Abul-Faraj A, Li L, Ghosh N, Piatek M, Mahjoub A, Aouida M, Piatek A, Baltes NJ, Voytas DF (2015) Efficient virus-mediated genome editing in plants using the CRISPR/Cas9 system. Mole Plant 8:1288–1291

Allard RW (1999) Principles of Plant Breeding. 2nd Edition, Wiley

Araki M, Ishii T (2015) Towards social acceptance of plant breeding by genome editing. Trends Plant Sci 20:145–149

Baltes NJ, Hummel AW, Konecna E, Cegan R, Bruns AN, Bisaro DM, Voytas DF (2015) Conferring resistance to geminiviruses with the CRISPR–Cas prokaryotic immune system. Nat Plants 1:15145

Bao A, Burritt DJ, Chen H, Zhou X, Cao D, Tran L-SP (2019a) The CRISPR/Cas9 system and its applications in crop genome editing. Crit Rev Biotechnol 39:321–336

Bao A, Chen H, Chen L, Chen S, Hao Q, Guo W, Qiu D, Shan Z, Yang Z, Yuan S, Zhang C, Zhang X, Liu B, Kong F, Li X, Zhou X, Tran LP, Cao D (2019b) CRISPR/Cas9-mediated targeted mutagenesis of *GmSPL9* genes alters plant architecture in soybean. BMC Plant Biol 19:131

Bari VK, Nassar JA, Kheredin SM, Gal-On A, Ron M, Britt A, Steele D, Yoder J, Aly R (2019) CRISPR/Cas9-mediated mutagenesis of *CAROTENOID CLEAVAGE DIOXYGENASE 8* in tomato provides resistance against the parasitic weed *Phelipanche aegyptiaca*. Sci Rep 9:11438

Barman HN, Sheng Z, Fiaz S, Zhong M, Wu Y, Cai Y, Wang W, Jiao G, Tang S, Wei X, Hu P (2019) Generation of a new thermo-sensitive genic male sterile rice line by targeted mutagenesis of *TMS5* gene through CRISPR/Cas9 system. BMC Plant Biol 19:109

Bawa A, Anilakumar K (2013) Genetically modified foods: safety, risks and public concerns—a review. J Food Sci Technol 50:1035–1046
Bevan MW, Flavell RB, Chilton M-D (1983) A chimaeric antibiotic resistance gene as a selectable marker for plant cell transformation. Nature 304:184
Bhowmik P, Ellison E, Polley B, Bollina V, Kulkarni M, Ghanbarnia K, Song H, Gao C, Voytas DF, Kagale S (2018) Targeted mutagenesis in wheat microspores using CRISPR/Cas9. Sci Rep 8:6502
Bilichak A, Gaudet D, Laurie J (2020) Emerging genome engineering tools in crop research and breeding. In: Cereal genomics. Springer, pp 165–181
Bogdanove AJ, Voytas DF (2011) TAL effectors: customizable proteins for DNA targeting. Science 333:1843–1846
Braatz J, Harloff H-J, Mascher M, Stein N, Himmelbach A, Jung C (2017) CRISPR-Cas9 targeted mutagenesis leads to simultaneous modification of different homoeologous gene copies in polyploid oilseed rape (*Brassica napus*). Plant Physiol 174:935–942
Brinegar K, Yetisen AK, Choi S, Vallillo E, Ruiz-Esparza GU, Prabhakar AM, Khademhosseini A, Yun S-H (2017) The commercialization of genome-editing technologies. Crit Rev Biotechnol 37:924–932
Brooks C, Nekrasov V, Lippman ZB, Van Eck J (2014) Efficient gene editing in tomato in the first generation using the clustered regularly interspaced short palindromic repeats/CRISPR-associated9 system. Plant Physiol 166:1292–1297
Brouns SJ, Jore MM, Lundgren M, Westra ER, Slijkhuis RJ, Snijders AP, Dickman MJ, Makarova KS, Koonin EV, Van Der Oost J (2008) Small CRISPR RNAs guide antiviral defense in prokaryotes. Science 321:960–964
Bruening G, Lyons J (2000) The case of the FLAVR SAVR tomato. Calif Agric 54:6–7
Butt H, Eid A, Ali Z, Atia MA, Mokhtar MM, Hassan N, Lee CM, Bao G, Mahfouz MM (2017) Efficient CRISPR/Cas9-mediated genome editing using a chimeric single-guide RNA molecule. Front Plant Sci 8:1441
Cai Y, Chen L, Liu X, Sun S, Wu C, Jiang B, Han T, Hou W (2015) CRISPR/Cas9-mediated genome editing in soybean hairy roots. PLoS ONE 10:e0136064
Cai Y, Chen L, Sun S, Wu C, Yao W, Jiang B, Han T, Hou W (2018a) CRISPR/Cas9-mediated deletion of large genomic fragments in soybean. Int J Mol Sci 19
Cai Y, Chen L, Sun S, Wu C, Yao W, Jiang B, Han T, Hou W (2018b) CRISPR/Cas9-mediated deletion of large genomic fragments in soybean. Int J Mol Sci 19:3835
Cai Y, Wang L, Chen L, Wu T, Liu L, Sun S, Wu C, Yao W, Jiang B, Yuan S, Han T, Hou W (2019) Mutagenesis of GmFT2a and GmFT5a mediated by CRISPR/Cas9 contributes for expanding the regional adaptability of soybean. Plant Biotechnol J 18:298–309
Cardi T, D'Agostino N, Tripodi P (2017) Genetic transformation and genomic resources for next-generation precise genome engineering in vegetable crops. Front Plant Sci 8:241
Carroll D (2011) Genome engineering with zinc-finger nucleases. Genetics 188:773–782
Čermák T, Baltes NJ, Čegan R, Zhang Y, Voytas DF (2015) High-frequency, precise modification of the tomato genome. Genome Biol 16:232
Chandrasekaran J, Brumin M, Wolf D, Leibman D, Klap C, Pearlsman M, Sherman A, Arazi T, Gal-On A (2016) Development of broad virus resistance in non-transgenic cucumber using CRISPR/Cas9 technology. Mole Plant Pathol 17:1140–1153
Char SN, Neelakandan AK, Nahampun H, Frame B, Main M, Spalding MH, Becraft PW, Meyers BC, Walbot V, Wang K, Yang B (2017) An Agrobacterium-delivered CRISPR/Cas9 system for high-frequency targeted mutagenesis in maize. Plant Biotechnol J 15:257–268
Chen R, Xu Q, Liu Y, Zhang J, Ren D, Wang G (2018) Generation of transgene-free maize male sterile lines using the CRISPR/Cas9 system. Front Plant Sci 9:1180
Chen S, Zhang N, Zhang Q, Zhou G, Tian H, Hussain S, Ahmed S, Wang T, Wang S (2019) Genome editing to integrate seed size and abiotic stress tolerance traits in Arabidopsis reveals a role for *DPA4* and *SOD7* in the regulation of inflorescence architecture. Int J Mol Sci 20:2695

Chilcoat D, Liu ZB, Sander J (2017) Use of CRISPR/Cas9 for Crop Improvement in Maize and Soybean. Prog Mol Biol Transl Sci 149:27–46

Cohen SN, Chang AC, Boyer HW, Helling RB (1973) Construction of biologically functional bacterial plasmids in vitro. Proc Natl Acad Sci 70:3240–3244

Cong L, Ran FA, Cox D, Lin S, Barretto R, Habib N, Hsu PD, Wu X, Jiang W, Marraffini LA (2013) Multiplex genome engineering using CRISPR/Cas systems. Science 339:819–823

D'Ambrosio C, Stigliani AL, Giorio G (2018) CRISPR/Cas9 editing of carotenoid genes in tomato. Transgenic Res 27:367–378

Do PT, Nguyen CX, Bui HT, Tran LTN, Stacey G, Gillman JD, Zhang ZJ, Stacey MG (2019) Demonstration of highly efficient dual gRNA CRISPR/Cas9 editing of the homeologous *GmFAD2-1A* and *GmFAD2-1B* genes to yield a high oleic, low linoleic and alpha-linolenic acid phenotype in soybean. BMC Plant Biol 19:311

Doll NM, Gilles LM, Gerentes MF, Richard C, Just J, Fierlej Y, Borrelli VMG, Gendrot G, Ingram GC, Rogowsky PM, Widiez T (2019) Single and multiple gene knockouts by CRISPR-Cas9 in maize. Plant Cell Rep 38:487–501

Du H, Zeng X, Zhao M, Cui X, Wang Q, Yang H, Cheng H, Yu D (2016) Efficient targeted mutagenesis in soybean by TALENs and CRISPR/Cas9. J Biotechnol 217:90–97

Eckerstorfer MF, Engelhard M, Heissenberger A, Simon S, Teichmann H (2019) Plants developed by new genetic modification techniques-comparison of existing regulatory frameworks in the EU and non-EU countries. Front Bioeng Biotechnol 7:26

Fan D, Liu T, Li C, Jiao B, Li S, Hou Y, Luo K (2015) Efficient CRISPR/Cas9-mediated targeted mutagenesis in *Populus* in the first generation. Sci Rep 5:12217

Fauser F, Schiml S, Puchta H (2014) Both CRISPR/CAS-based nucleases and nickases can be used efficiently for genome engineering in *Arabidopsis thaliana*. Plant J 79:348–359

Feng Z, Zhang B, Ding W, Liu X, Yang D-L, Wei P, Cao F, Zhu S, Zhang F, Mao Y (2013) Efficient genome editing in plants using a CRISPR/Cas system. Cell Res 23:1229

Feng Z, Mao Y, Xu N, Zhang B, Wei P, Yang D-L, Wang Z, Zhang Z, Zheng R, Yang L (2014) Multigeneration analysis reveals the inheritance, specificity, and patterns of CRISPR/Cas-induced gene modifications in *Arabidopsis*. Proc Natl Acad Sci 111:4632–4637

Feng C, Yuan J, Wang R, Liu Y, Birchler JA, Han F (2016) Efficient Targeted Genome Modification in Maize Using CRISPR/Cas9 System. J Genet Genomics 43:37–43

Feng C, Su H, Bai H, Wang R, Liu Y, Guo X, Liu C, Zhang J, Yuan J, Birchler JA, Han F (2018) High-efficiency genome editing using a *dmc1* promoter-controlled CRISPR/Cas9 system in maize. Plant Biotechnol J 16:1848–1857

Filler Hayut S, Melamed Bessudo C, Levy AA (2017) Targeted recombination between homologous chromosomes for precise breeding in tomato. Nat Commun 8:15605

Fister AS, Landherr L, Maximova SN, Guiltinan MJ (2018) Transient expression of CRISPR/Cas9 machinery targeting TcNPR3 enhances defense response in *Theobroma cacao*. Front Plant Sci 9:268

Funk C, Rainie L (2015) Public and scientists' views on science and society. Pew Research Center 29

Gomez MA, Lin ZD, Moll T, Chauhan RD, Hayden L, Renninger K, Beyene G, Taylor NJ, Carrington JC, Staskawicz BJ (2019) Simultaneous CRISPR/Cas9-mediated editing of cassava eIF 4E isoforms nCBP-1 and nCBP-2 reduces cassava brown streak disease symptom severity and incidence. Plant Biotechnol J 17:421–434

Gupta RM, Musunuru K (2014) Expanding the genetic editing tool kit: ZFNs, TALENs, and CRISPR-Cas9. J Clin Investig 124:4154–4161

Gupta PK, Tsuchiya T (1991) Chromosome engineering in plants: genetics, breeding, evolution. Elsevier Science

Hensel G, Pouramini P, Hiekel S, Reuter P, Baier S, Kumlehn J (2018) Generation of new barley mutant alleles of *LIPOXYGENASE 1* using CRISPR RNA/Cas9-endonuclease technology. In: In vitro cellular & developmental biology-plant, pp S87–S88

Hu B, Li D, Liu X, Qi J, Gao D, Zhao S, Huang S, Sun J, Yang L (2017) Engineering non-transgenic gynoecious cucumber using an improved transformation protocol and optimized CRISPR/Cas9 system. Mole Plant 10:1575–1578
Huang S, Weigel D, Beachy RN, Li J (2016) A proposed regulatory framework for genome-edited crops. Nat Genet 48:109
Huang X, Zeng X, Li J, Zhao D (2017) Construction and analysis of *tify1a* and *tify1b* mutants in rice (*Oryza sativa*) based on CRISPR/Cas9 technology. J Agri Biotechnol 25:1003–1012
Hyun Y, Kim J, Cho SW, Choi Y, Kim J-S, Coupland G (2015) Site-directed mutagenesis in *Arabidopsis thaliana* using dividing tissue-targeted RGEN of the CRISPR/Cas system to generate heritable null alleles. Planta 241:271–284
Iqbal Z, Sattar MN, Shafiq M (2016) CRISPR/Cas9: a tool to circumscribe cotton leaf curl disease. Front Plant Sci 7:475
ISAAA (2017) Global status of commercialized biotech/GM crops in 2017: biotech crop adoption surges as economic benefits accumulate in 22 years. https://www.isaaa.org
Ishii T, Araki M (2017) A future scenario of the global regulatory landscape regarding genome-edited crops. GM Crops Food 8:44–56
Ito Y, Nishizawa-Yokoi A, Endo M, Mikami M, Toki S (2015) CRISPR/Cas9-mediated mutagenesis of the RIN locus that regulates tomato fruit ripening. Biochem Biophys Res Commun 467:76–82
Jacobs TB, LaFayette PR, Schmitz RJ, Parrott WA (2015) Targeted genome modifications in soybean with CRISPR/Cas9. BMC Biotechnol 15:16
Jaenisch R, Mintz B (1974) Simian virus 40 DNA sequences in DNA of healthy adult mice derived from preimplantation blastocysts injected with viral DNA. Proc Natl Acad Sci 71:1250–1254
Ji X, Zhang H, Zhang Y, Wang Y, Gao C (2015) Establishing a CRISPR–Cas-like immune system conferring DNA virus resistance in plants. Nat Plants 1:15144
Jia H, Wang N (2014) Targeted genome editing of sweet orange using Cas9/sgRNA. PLoS ONE 9:e93806
Jia H, Zhang Y, Orbović V, Xu J, White FF, Jones JB, Wang N (2017) Genome editing of the disease susceptibility gene *CsLOB1* in citrus confers resistance to citrus canker. Plant Biotechnol J 15:817–823
Jiang W, Zhou H, Bi H, Fromm M, Yang B, Weeks DP (2013) Demonstration of CRISPR/Cas9/sgRNA-mediated targeted gene modification in Arabidopsis, tobacco, sorghum and rice. Nucleic Acids Res 41:e188–e188
Jiang W, Yang B, Weeks DP (2014) Efficient CRISPR/Cas9-mediated gene editing in Arabidopsis thaliana and inheritance of modified genes in the T2 and T3 generations. PloS one 9(6):e99225
Jones L (1999) Genetically modified foods. BMJ 318:581–584
Jones HD (2015) Regulatory uncertainty over genome editing. Nat Plants 1(10):1038
Kalaitzandonakes N, Kaufman J, Miller D (2014) Potential economic impacts of zero thresholds for unapproved GMOs: the EU case. Food Policy 45:146–157
Kanchiswamy CN, Malnoy M, Velasco R, Kim J-S, Viola R (2015) Non-GMO genetically edited crop plants. Trends Biotechnol 33:489–491
Kim J, Kim J-S (2016) Bypassing GMO regulations with CRISPR gene editing. Nat Biotechnol 34:1014
Kim D, Alptekin B, Budak H (2018) CRISPR/Cas9 genome editing in wheat. Funct Integr Genom 18:31–41
Klümper W, Qaim M (2014) A meta-analysis of the impacts of genetically modified crops. PLoS ONE 9:e111629
Kramer MG, Redenbaugh K (1994) Commercialization of a tomato with an antisense polygalacturonase gene: The FLAVR SAVR™ tomato story. Euphytica 79:293–297
Kumar N, Galli M, Ordon J, Stuttmann J, Kogel KH, Imani J (2018) Further analysis of barley MORC 1 using a highly efficient RNA-guided Cas9 gene-editing system. Plant Biotechnol J 16:1892–1903
Langner T, Kamoun S, Belhaj K (2018) CRISPR crops: plant genome editing toward disease resistance. Annu Rev Phytopathol 56:479–512

Larkin PJ, Scowcroft WR (1981) Somaclonal variation—a novel source of variability from cell cultures for plant improvement. Theoret Appl Genet 60:197–214

Lawrenson T, Shorinola O, Stacey N, Li C, Østergaard L, Patron N, Uauy C, Harwood W (2015) Induction of targeted, heritable mutations in barley and *Brassica oleracea* using RNA-guided Cas9 nuclease. Genome Biol 16:258

Lee K, Eggenberger AL, Banakar R, McCaw ME, Zhu H, Main M, Kang M, Gelvin SB, Wang K (2019a) CRISPR/Cas9-mediated targeted T-DNA integration in rice. Plant Mol Biol 99:317–328

Lee K, Zhang Y, Kleinstiver BP, Guo JA, Aryee MJ, Miller J, Malzahn A, Zarecor S, Lawrence-Dill CJ, Joung JK, Qi Y, Wang K (2019b) Activities and specificities of CRISPR/Cas9 and Cas12a nucleases for targeted mutagenesis in maize. Plant Biotechnol J 17:362–372

Lee K, Zhu H, Yang B, Wang K (2019c) An Agrobacterium-mediated CRISPR/Cas9 platform for genome editing in maize. Methods Mol Biol 1917:121–143

Li J-F, Norville JE, Aach J, McCormack M, Zhang D, Bush J, Church GM, Sheen J (2013) Multiplex and homologous recombination–mediated genome editing in *Arabidopsis* and *Nicotiana benthamiana* using guide RNA and Cas9. Nat Biotechnol 31:688

Li Z, Liu Z-B, Xing A, Moon BP, Koellhoffer JP, Huang L, Ward RT, Clifton E, Falco SC, Cigan AM (2015) Cas9-guide RNA directed genome editing in soybean. Plant Physiol 169:960–970

Li M, Li X, Zhou Z, Wu P, Fang M, Pan X, Lin Q, Luo W, Wu G, Li H (2016) Reassessment of the four yield-related genes *Gn1a, DEP1, GS3,* and*IPA1* in rice using a CRISPR/Cas9 system. Front Plant Sci 7:377

Li C, Liu C, Qi X, Wu Y, Fei X, Mao L, Cheng B, Li X, Xie C (2017a) RNA-guided Cas9 as an in vivo desired-target mutator in maize. Plant Biotechnol J 15:1566–1576

Li X, Zhou W, Ren Y, Tian X, Lv T, Wang Z, Fang J, Chu C, Yang J, Bu Q (2017b) High-efficiency breeding of early-maturing rice cultivars via CRISPR/Cas9-mediated genome editing. J Genet Genom 44:175

Li HQ, Chen C, Chen RR, Song XW, Li JN, Zhu YM, Ding XD (2018a) Preliminary analysis of the role of GmSnRK1.1 and GmSnRK1.2 in the ABA and alkaline stress response of the soybean using the CRISPR/Cas9-based gene double-knockout system. Yi Chuan 40:496–507

Li R, Fu D, Zhu B, Luo Y, Zhu H (2018b) CRISPR/Cas9-mediated mutagenesis of lncRNA1459 alters tomato fruit ripening. Plant J 94:513–524

Li R, Zhang L, Wang L, Chen L, Zhao R, Sheng J, Shen L (2018c) Reduction of tomato-plant chilling tolerance by CRISPR-Cas9-mediated *SlCBF1* mutagenesis. J Agric Food Chem 66:9042–9051

Li X, Wang Y, Chen S, Tian H, Fu D, Zhu B, Luo Y, Zhu H (2018d) Lycopene is enriched in tomato fruit by CRISPR/Cas9-mediated multiplex genome editing. Front Plant Sci 9:559

Li C, Nguyen V, Liu J, Fu W, Chen C, Yu K, Cui Y (2019a) Mutagenesis of seed storage protein genes in Soybean using CRISPR/Cas9. BMC Res Notes 12:176

Li R, Liu C, Zhao R, Wang L, Chen L, Yu W, Zhang S, Sheng J, Shen L (2019b) CRISPR/Cas9-Mediated SlNPR1 mutagenesis reduces tomato plant drought tolerance. BMC Plant Biol 19:38

Liang Z, Zhang K, Chen K, Gao C (2014) Targeted mutagenesis in Zea mays using TALENs and the CRISPR/Cas system. J Genet Genom 41:63–68

Liu J, Gunapati S, Mihelich NT, Stec AO, Michno JM, Stupar RM (2019) Genome editing in soybean with CRISPR/Cas9. Methods Mol Biol 1917:217–234

Lu Y, Zhu J-K (2017) Precise editing of a target base in the rice genome using a modified CRISPR/Cas9 system. Mole Plant 10:523–525

Lu HP, Liu SM, Xu SL, Chen WY, Zhou X, Tan YY, Huang JZ, Shu QY (2017) CRISPR-S: an active interference element for a rapid and inexpensive selection of genome-edited, transgene-free rice plants. Plant Biotechnol J 15:1371

Lucht J (2015) Public acceptance of plant biotechnology and GM crops. Viruses 7:4254–4281

Lusser M, Parisi C, Plan D, Rodríguez-Cerezo E (2011) New plant breeding techniques: state-of-the-art and prospects for commercial development. Publications Office of the European Union

Ma JK, Drake PM, Christou P (2003) Genetic modification: the production of recombinant pharmaceutical proteins in plants. Nat Rev Genet 4:794

Ma X, Zhang Q, Zhu Q, Liu W, Chen Y, Qiu R, Wang B, Yang Z, Li H, Lin Y (2015) A robust CRISPR/Cas9 system for convenient, high-efficiency multiplex genome editing in monocot and dicot plants. Mol Plant 8(8):1274–1284

Mali P, Yang L, Esvelt KM, Aach J, Guell M, DiCarlo JE, Norville JE, Church GM (2013) RNA-guided human genome engineering via Cas9. Science 339:823–826

Maluszynski M (2001) Officially released mutant varieties–the FAO/IAEA database. Plant Cell Tissue Organ Cult 65:175–177

Malzahn AA, Tang X, Lee K, Ren Q, Sretenovic S, Zhang Y, Chen H, Kang M, Bao Y, Zheng X, Deng K, Zhang T, Salcedo V, Wang K, Qi Y (2019) Application of CRISPR-Cas12a temperature sensitivity for improved genome editing in rice, maize, and *Arabidopsis*. BMC Biol 17:9

Marraffini LA, Sontheimer EJ (2008) CRISPR interference limits horizontal gene transfer in staphylococci by targeting DNA. Science 322:1843–1845

Mehta D, Stürchler A, Anjanappa RB, Zaidi SS-e-A, Hirsch-Hoffmann M, Gruissem W, Vanderschuren H (2019) Linking CRISPR-Cas9 interference in cassava to the evolution of editing-resistant geminiviruses. Genome Biol 20:80

Miao J, Guo D, Zhang J, Huang Q, Qin G, Zhang X, Wan J, Gu H, Qu L-J (2013) Targeted mutagenesis in rice using CRISPR-Cas system. Cell Res 23:1233

Miao C, Xiao L, Hua K, Zou C, Zhao Y, Bressan RA, Zhu JK (2018) Mutations in a subfamily of abscisic acid receptor genes promote rice growth and productivity. Proc Natl Acad Sci USA 115(23):6058–6063

Michno JM, Wang X, Liu J, Curtin SJ, Kono TJ, Stupar RM (2015) CRISPR/Cas mutagenesis of soybean and Medicago truncatula using a new web-tool and a modified Cas9 enzyme. GM Crops Food 6:243–252

Mohanta TK, Bashir T, Hashem A, Abd Allah EF, Bae H (2017) Genome editing tools in plants. Genes 8:399

Mojica FJ, García-Martínez J, Soria E (2005) Intervening sequences of regularly spaced prokaryotic repeats derive from foreign genetic elements. J Mol Evol 60:174–182

Nekrasov V, Staskawicz B, Weigel D, Jones JD, Kamoun S (2013) Targeted mutagenesis in the model plant *Nicotiana benthamiana* using Cas9 RNA-guided endonuclease. Nat Biotechnol 31:691

Nekrasov V, Wang C, Win J, Lanz C, Weigel D, Kamoun S (2017) Rapid generation of a transgene-free powdery mildew resistant tomato by genome deletion. Sci Rep 7:482

Nieves-Cordones M, Mohamed S, Tanoi K, Kobayashi NI, Takagi K, Vernet A, Guiderdoni E, Périn C, Sentenac H, Véry AA (2017) Production of low-Cs+ rice plants by inactivation of the K+ transporter *OsHAK1* with the CRISPR-Cas system. Plant J 92:43–56

Nonaka S, Arai C, Takayama M, Matsukura C, Ezura H (2017) Efficient increase of γ-aminobutyric acid (GABA) content in tomato fruits by targeted mutagenesis. Sci Rep 7:7057

Odipio J, Alicai T, Ingelbrecht I, Nusinow DA, Bart R, Taylor NJ (2017) Efficient CRISPR/Cas9 genome editing of phytoene desaturase in cassava. Front Plant Sci 8:1780

O'Keefe M, Perrault S, Halpern J, Ikemoto L, Yarborough M (2015) "Editing" genes: a case study about how language matters in bioethics. Am J Bioethics 15:3–10

Okuzaki A, Ogawa T, Koizuka C, Kaneko K, Inaba M, Imamura J, Koizuka N (2018) CRISPR/Cas9-mediated genome editing of the fatty acid desaturase 2 gene in Brassica napus. Plant Physiol Biochem 131:63–69

Ortigosa A, Gimenez-Ibanez S, Leonhardt N, Solano R (2019) Design of a bacterial speck resistant tomato by CRISPR/Cas9-mediated editing of *SlJAZ2*. Plant Biotechnol J 17:665–673

Paixão JFR, Gillet F-X, Ribeiro TP, Bournaud C, Lourenço-Tessutti IT, Noriega DD, de Melo BP, de Almeida-Engler J, Grossi-de-Sa MF (2019) Improved drought stress tolerance in Arabidopsis by CRISPR/dCas9 fusion with a Histone AcetylTransferase. Sci Rep 9:8080

Pan C, Ye L, Qin L, Liu X, He Y, Wang J, Chen L, Lu G (2016) CRISPR/Cas9-mediated efficient and heritable targeted mutagenesis in tomato plants in the first and later generations. Sci Rep 6:24765

Parekh SR (2004) The GMO handbook: genetically modified animals, microbes, and plants in biotechnology. Springer Science & Business Media

Park J-J, Yoo CG, Flanagan A, Pu Y, Debnath S, Ge Y, Ragauskas AJ, Wang Z-Y (2017) Defined tetra-allelic gene disruption of the 4-coumarate: coenzyme A ligase 1 (Pv4CL1) gene by CRISPR/Cas9 in switchgrass results in lignin reduction and improved sugar release. Biotechnol Biofuels 10:284

Parkhi V, Bhattacharya A, Choudhary S, Pathak R, Gawade V, Palan B, Alamalakala L, Mikkilineni V, Char B (2018) Demonstration of CRISPR-cas9-mediated pds gene editing in a tomato hybrid parental line. Indian J Genet 78:132–137

Peng A, Chen S, Lei T, Xu L, He Y, Wu L, Yao L, Zou X (2017) Engineering canker-resistant plants through CRISPR/Cas9-targeted editing of the susceptibility gene CsLOB1 promoter in citrus. Plant Biotechnol J 15:1509–1519. https://doi.org/10.1111/pbi.12733

Petolino JF, Srivastava V, Daniell H (2016) Editing Plant Genomes: a new era of crop improvement. Plant Biotechnol J 14:435–436

Podevin N, Davies HV, Hartung F, Nogue F, Casacuberta JM (2013) Site-directed nucleases: a paradigm shift in predictable, knowledge-based plant breeding. Trends Biotechnol 31:375–383

Poehlman JM (2013) Breeding field crops. Springer Science & Business Media

Pourcel C, Salvignol G, Vergnaud G (2005) CRISPR elements in *Yersinia pestis* acquire new repeats by preferential uptake of bacteriophage DNA, and provide additional tools for evolutionary studies. Microbiology 151:653–663

Pyott DE, Sheehan E, Molnar A (2016) Engineering of CRISPR/Cas9-mediated potyvirus resistance in transgene-free Arabidopsis plants. Mole Plant Pathol 17:1276–1288

Qi W, Zhu T, Tian Z, Li C, Zhang W, Song R (2016) High-efficiency CRISPR/Cas9 multiplex gene editing using the glycine tRNA-processing system-based strategy in maize. BMC Biotechnol 16:58

Ricroch AE, Guillaume-Hofnung M, Kuntz M (2018) The ethical concerns about transgenic crops. Portland Press Limited

Rodríguez-Leal D, Lemmon ZH, Man J, Bartlett ME, Lippman ZB (2017) Engineering quantitative trait variation for crop improvement by genome editing. Cell 171:470–480

Ryan CD, Smyth SJ (2012) Economic implications of low-level presence in a zero-tolerance European import market: the case of Canadian Triffid flax. AgBioForum 15(1):21-30

Saika H, Mori A, Endo M, Toki S (2019) Targeted deletion of rice retrotransposon Tos17 via CRISPR/Cas9. Plant Cell Rep 38:455–458

Sánchez-León S, Gil-Humanes J, Ozuna CV, Giménez MJ, Sousa C, Voytas DF, Barro F (2018) Low-gluten, nontransgenic wheat engineered with CRISPR/Cas9. Plant Biotechnol J 16:902–910

Sapranauskas R, Gasiunas G, Fremaux C, Barrangou R, Horvath P, Siksnys V (2011) The *Streptococcus thermophilus* CRISPR/Cas system provides immunity in *Escherichia coli*. Nucleic Acids Res 39:9275–9282

Schiml S, Fauser F, Puchta H (2014) The CRISPR/CAS as system can be used as nuclease for in planta gene targeting and as paired nickases for directed mutagenesis in *Arabidopsis* resulting in heritable progeny. Plant J 80:1139–1150

Schindele A, Dorn A, Puchta H (2020) CRISPR/Cas brings plant biology and breeding into the fast lane. Curr Opin Biotechnol 61:7–14

Shan Q, Wang Y, Li J, Zhang Y, Chen K, Liang Z, Zhang K, Liu J, Xi JJ, Qiu J-L (2013) Targeted genome modification of crop plants using a CRISPR-Cas system. Nat Biotechnol 31:686

Shan Q, Wang Y, Li J, Gao C (2014) Genome editing in rice and wheat using the CRISPR/Cas system. Nat Protoc 9:2395

Shen R, Wang L, Liu X, Wu J, Jin W, Zhao X, Xie X, Zhu Q, Tang H, Li Q (2017) Genomic structural variation-mediated allelic suppression causes hybrid male sterility in rice. Nat Commun 8:1310

Shi J, Gao H, Wang H, Lafitte HR, Archibald RL, Yang M, Hakimi SM, Mo H, Habben JE (2017) ARGOS8 variants generated by CRISPR-Cas9 improve maize grain yield under field drought stress conditions. Plant Biotechnol J 15:207–216

Shimatani Z, Kashojiya S, Takayama M, Terada R, Arazoe T, Ishii H, Teramura H, Yamamoto T, Komatsu H, Miura K, Ezura H, Nishida K, Ariizumi T, Kondo A (2017) Targeted base editing in rice and tomato using a CRISPR-Cas9 cytidine deaminase fusion. Nat Biotechnol 35:441–443

Shukla VK, Doyon Y, Miller JC, DeKelver RC, Moehle EA, Worden SE, Mitchell JC, Arnold NL, Gopalan S, Meng X (2009) Precise genome modification in the crop species *Zea mays* using zinc-finger nucleases. Nature 459:437

Smyth SJ (2017) Genetically modified crops, regulatory delays, and international trade. Food Energy Secur 6:78–86

Smyth SJ, Lassoued R (2019) Agriculture R&D implications of the CJEU's gene-specific mutagenesis ruling. Trends Biotechnol 37:337–340

Soyk S, Müller NA, Park SJ, Schmalenbach I, Jiang K, Hayama R, Zhang L, Van Eck J, Jiménez-Gómez JM, Lippman ZB (2017) Variation in the flowering gene *SELF PRUNING 5G* promotes day-neutrality and early yield in tomato. Nat Genet 49:162

Sprink T, Metje J, Hartung F (2015) Plant genome editing by novel tools: TALEN and other sequence specific nucleases. Curr Opin Biotechnol 32:47–53

Srivastava S, Rahman MH, Shah S, Kav NN (2006) Constitutive expression of the pea ABA-responsive 17 (ABR17) cDNA confers multiple stress tolerance in *Arabidopsis thaliana.* Plant Biotechnol J 4:529–549

Sun X, Hu Z, Chen R, Jiang Q, Song G, Zhang H, Xi Y (2015) Targeted mutagenesis in soybean using the CRISPR-Cas9 system. Sci Rep 5:10342

Sun Y, Zhang X, Wu C, He Y, Ma Y, Hou H, Guo X, Du W, Zhao Y, Xia L (2016) Engineering herbicide-resistant rice plants through CRISPR/Cas9-mediated homologous recombination of acetolactate synthase. Mole Plant 9:628–631

Sun Y, Jiao G, Liu Z, Zhang X, Li J, Guo X, Du W, Du J, Francis F, Zhao Y (2017) Generation of high-amylose rice through CRISPR/Cas9-mediated targeted mutagenesis of starch branching enzymes. Front Plant Sci 8:298

Suprasanna P, Mirajkar S, Bhagwat S (2015) Induced mutations and crop improvement. In: Plant biology and biotechnology. Springer, pp 593–617

Svitashev S, Schwartz C, Lenderts B, Young JK, Mark Cigan A (2016) Genome editing in maize directed by CRISPR-Cas9 ribonucleoprotein complexes. Nat Commun 7:13274

Tang L, Mao B, Li Y, Lv Q, Zhang L, Chen C, He H, Wang W, Zeng X, Shao Y (2017) Knockout of *OsNramp5* using the CRISPR/Cas9 system produces low Cd-accumulating *indica* rice without compromising yield. Sci Rep 7:14438

Tashkandi M, Ali Z, Aljedaani F, Shami A, Mahfouz MM (2018) Engineering resistance against Tomato yellow leaf curl virus via the CRISPR/Cas9 system in tomato. Plant Signal Behav 13:e1525996

Taylor A (2019) Companies use CRISPR to improve crops. Bio Business, The Scientist

Tian S, Jiang L, Cui X, Zhang J, Guo S, Li M, Zhang H, Ren Y, Gong G, Zong M (2018) Engineering herbicide-resistant watermelon variety through CRISPR/Cas9-mediated base-editing. Plant Cell Rep 37:1353–1356

Tomlinson L, Yang Y, Emenecker R, Smoker M, Taylor J, Perkins S, Smith J, MacLean D, Olszewski NE, Jones JD (2019) Using CRISPR/Cas9 genome editing in tomato to create a gibberellin-responsive dominant dwarf DELLA allele. Plant Biotechnol J 17:132–140

Tourte Y (2019) Genetic engineering and biotechnology: concepts, methods and agronomic applications. CRC Press

Townsend JA, Wright DA, Winfrey RJ, Fu F, Maeder ML, Joung JK, Voytas DF (2009) High-frequency modification of plant genes using engineered zinc-finger nucleases. Nature 459:442

Ueta R, Abe C, Watanabe T, Sugano SS, Ishihara R, Ezura H, Osakabe Y, Osakabe K (2017) Rapid breeding of parthenocarpic tomato plants using CRISPR/Cas9. Sci Rep 7:507

Urnov FD, Rebar EJ, Holmes MC, Zhang HS, Gregory PD (2010) Genome editing with engineered zinc finger nucleases. Nat Rev Genet 11:636

Uzogara SG (2000) The impact of genetic modification of human foods in the 21st century: a review. Biotechnol Adv 18:179–206

Valin H, Sands RD, Van der Mensbrugghe D, Nelson GC, Ahammad H, Blanc E, Bodirsky B, Fujimori S, Hasegawa T, Havlik P (2014) The future of food demand: understanding differences in global economic models. Agri Econ 45:51–67 van Harten AM (1998) Mutation breeding: theory and practical applications. Cambridge University Press

Veillet F, Perrot L, Chauvin L, Kermarrec MP, Guyon-Debast A, Chauvin JE, Nogue F, Mazier M (2019) Transgene-free genome editing in tomato and potato plants using agrobacterium-mediated delivery of a CRISPR/Cas9 cytidine base editor. Int J Mol Sci 20

Wang Y, Cheng X, Shan Q, Zhang Y, Liu J, Gao C, Qiu J-L (2014) Simultaneous editing of three homoeoalleles in hexaploid bread wheat confers heritable resistance to powdery mildew. Nat Biotechnol 32:947

Wang F, Wang C, Liu P, Lei C, Hao W, Gao Y, Liu Y-G, Zhao K (2016) Enhanced rice blast resistance by CRISPR/Cas9-targeted mutagenesis of the ERF transcription factor gene *OsERF922*. PLoS ONE 11:e0154027

Wang L, Chen L, Li R, Zhao R, Yang M, Sheng J, Shen L (2017) Reduced drought tolerance by CRISPR/Cas9-mediated *SlMAPK3* mutagenesis in tomato plants. J Agri Food Chem 65:8674–8682

Wang P, Zhang J, Sun L, Ma Y, Xu J, Liang S, Deng J, Tan J, Zhang Q, Tu L (2018a) High efficient multisites genome editing in allotetraploid cotton (*Gossypium hirsutum*) using CRISPR/Cas9 system. Plant Biotechnol J 16:137–150

Wang X, Tu M, Wang D, Liu J, Li Y, Li Z, Wang Y, Wang X (2018b) CRISPR/Cas9-mediated efficient targeted mutagenesis in grape in the first generation. Plant Biotechnol J 16:844–855

Wang D, Samsulrizal NH, Yan C, Allcock NS, Craigon J, Blanco-Ulate B, Ortega-Salazar I, Marcus SE, Bagheri HM, Perez Fons L, Fraser PD, Foster T, Fray R, Knox JP, Seymour GB (2019a) Characterization of CRISPR mutants targeting genes modulating pectin degradation in ripening tomato. Plant Physiol 179:544–557

Wang R, Tavano E, Lammers M, Martinelli AP, Angenent GC, de Maagd RA (2019b) Re-evaluation of transcription factor function in tomato fruit development and ripening with CRISPR/Cas9-mutagenesis. Sci Rep 9:1696

Wasmer M (2019) Roads forward for European GMO policy—uncertainties in wake of ECJ judgment have to be mitigated by regulatory reform. Front Bioeng Biotechnol 7

Wilson SA, Roberts SC (2014) Metabolic engineering approaches for production of biochemicals in food and medicinal plants. Curr Opin Biotechnol 26:174–182

Wolt JD, Wolf C (2018) Policy and governance perspectives for regulation of genome edited crops in the United States. Front Plant Sci 9

Wolt JD, Wang K, Yang B (2016) The regulatory status of genome edited crops. Plant Biotechnol J 14:510–518

Xie K, Yang Y (2013) RNA-guided genome editing in plants using a CRISPR–Cas system. Mole Plant 6:1975–1983

Xie K, Minkenberg B, Yang Y (2015) Boosting CRISPR/Cas9 multiplex editing capability with the endogenous tRNA-processing system. Proc Natl Acad Sci 112:3570–3575

Xie Y, Niu B, Long Y, Li G, Tang J, Zhang Y, Ren D, Liu YG, Chen L (2017) Suppression or knockout of *SaF/SaM* overcomes the *Sa*-mediated hybrid male sterility in rice. J Integr Plant Biol 59:669–679

Xu R, Li H, Qin R, Wang L, Li L, Wei P, Yang J (2014) Gene targeting using the *Agrobacterium tumefaciens*-mediated CRISPR-Cas system in rice. Rice 7:5

Xu R-F, Li H, Qin R-Y, Li J, Qiu C-H, Yang Y-C, Ma H, Li L, Wei P-C, Yang J-B (2015) Generation of inheritable and "transgene clean" targeted genome-modified rice in later generations using the CRISPR/Cas9 system. Sci Rep 5:11491

Yang H, Wu J-J, Tang T, Liu K-D, Dai C (2017) CRISPR/Cas9-mediated genome editing efficiently creates specific mutations at multiple loci using one sgRNA in Brassica napus. Sci Rep 7:7489

Yang Y, Zhu K, Li H, Han S, Meng Q, Khan SU, Fan C, Xie K, Zhou Y (2018) Precise editing of *CLAVATA* genes in *Brassica napus* L. regulates multilocular silique development. Plant Biotechnol J 16:1322–1335

Yu QH, Wang B, Li N, Tang Y, Yang S, Yang T, Xu J, Guo C, Yan P, Wang Q, Asmutola P (2017) CRISPR/Cas9-induced targeted mutagenesis and gene replacement to generate long-shelf life tomato lines. Sci Rep 7:11874

Zaidi SS-e-A, Tashkandi M, Mansoor S, Mahfouz MM (2016) Engineering plant immunity: using CRISPR/Cas9 to generate virus resistance. Front Plant Sci 7

Zhang H, Zhang J, Wei P, Zhang B, Gou F, Feng Z, Mao Y, Yang L, Zhang H, Xu N (2014) The CRISPR/Cas9 system produces specific and homozygous targeted gene editing in rice in one generation. Plant Biotechnol J 12:797–807

Zhang Y, Bai Y, Wu G, Zou S, Chen Y, Gao C, Tang D (2017) Simultaneous modification of three homoeologs of *TaEDR1* by genome editing enhances powdery mildew resistance in wheat. Plant J 91:714–724

Zhang J, Zhang H, Botella JR, Zhu JK (2018a) Generation of new glutinous rice by CRISPR/Cas9-targeted mutagenesis of the *Waxy* gene in elite rice varieties. J Integr Plant Biol 60:369–375

Zhang Y, Massel K, Godwin ID, Gao C (2018b) Applications and potential of genome editing in crop improvement. Genome Biol 19:210

Zhou H, Liu B, Weeks DP, Spalding MH, Yang B (2014) Large chromosomal deletions and heritable small genetic changes induced by CRISPR/Cas9 in rice. Nucleic Acids Res 42:10903–10914

Zhou J, Peng Z, Long J, Sosso D, Liu B, Eom JS, Huang S, Liu S, Vera Cruz C, Frommer WB (2015) Gene targeting by the TAL effector *PthXo2* reveals cryptic resistance gene for bacterial blight of rice. Plant J 82:632–643

Zhou X, Jacobs TB, Xue LJ, Harding SA, Tsai CJ (2015) Exploiting SNPs for biallelic CRISPR mutations in the outcrossing woody perennial *Populus* reveals 4-coumarate: CoA ligase specificity and redundancy. New Phytol 208:298–301

Zhou H, He M, Li J, Chen L, Huang Z, Zheng S, Zhu L, Ni E, Jiang D, Zhao B (2016) Development of commercial thermo-sensitive genic male sterile rice accelerates hybrid rice breeding using the CRISPR/Cas9-mediated *TMS5* editing system. Sci Rep 6:37395

Zhu C, Naqvi S, Gomez-Galera S, Pelacho AM, Capell T, Christou P (2007) Transgenic strategies for the nutritional enhancement of plants. Trends Plant Sci 12:548–555

Zhu J, Song N, Sun S, Yang W, Zhao H, Song W, Lai J (2016) Efficiency and inheritance of targeted mutagenesis in maize using CRISPR-Cas9. J Genet Genomics 43:25–36

Zimny T, Sowa S, Tyczewska A, Twardowski T (2019) Certain new plant breeding techniques and their marketability in the context of EU GMO legislation—recent developments. New Biotechnol 51:49–56

Zong Y, Wang Y, Li C, Zhang R, Chen K, Ran Y, Qiu J-L, Wang D, Gao C (2017) Precise base editing in rice, wheat and maize with a Cas9-cytidine deaminase fusion. Nat Biotechnol 35:438

Chapter 8
Translational Research Using CRISPR/Cas

Anshika Tyagi, Sandhya Sharma, Sanskriti Vats, Sajad Ali, Sandeep Kumar, Naveed Gulzar, and Ruspesh Deshmukh

Abstract Gene editing technologies have revolutionized the development of new and innovative bio-products in plants as well as animals in recent years. Of the different techniques available, CRISPR/Cas has been and will remain the basis of genome engineering for researchers across the world because of its easiness to use and specificity. There have been exciting milestones in the interpretation of plant biology, development of new and improved plant varieties in terms of quality, yield, and biotic and abiotic stress resistance using CRISPR technology. Besides, the identification of new characteristics, expanding the range of currently known traits and faster rate of trait development has been made possible through this technology. In this chapter, we have discussed about strategies focusing on the use of CRISPR/Cas for trait development and improvement in plants, regulation of growth and flowering, improving the nutritional value and strengthening the plant adaptation to various abiotic and biotic stresses. Further, we have also focused on improvements in the targeted activity of CRISPR/Cas system which has resulted in undertaking precise genetic modifications, generating specific knock-ins and knock-outs, and multiple changes in the genome simultaneously with negligible to no off-target effects.

Keywords Crispr/Cas tools · Applications · Biotic stress tolerance · Quality · Nutrition

---

A. Tyagi · S. Sharma
ICAR-National Institute for Plant Biotechnology, New Delhi, India

S. Vats · R. Deshmukh (✉)
National Agri-food Biotechnology Institute, Mohali, India
e-mail: rupesh0deshmukh@gmail.com

S. Ali · N. Gulzar
Center of Research for Development, University of Kashmir, Srinagar, India

S. Kumar
Xcelris Labs Ltd., Ahmedabad, India

A. Bhattacharya et al. (eds.), *CRISPR/Cas Genome Editing*, Concepts and Strategies in Plant Sciences, https://doi.org/10.1007/978-3-030-42022-2_8

8.1 Introduction

Since its discovery, CRISPR/Cas have been the most commonly availed gene-editing technology in research labs across the world. From its fundamental role as an adaptive immune system in microbes to its development as pioneering gene-editing tool, the technology has seen manifold growth in its use and development of variants with enhanced target specificity. It is more comfortable and less complex than its contemporaries like Transcription-Activator Like Effector Nucleases (TALENs) and Zinc finger nucleases (ZFNs), since TALENs and ZFNs require designing whole proteins whereas simple cloning practices can achieve editing via CRISPR/Cas. The basic structure of a CRISPR locus consists of short repetitive DNA sequences separated by spacer DNA and Cas is an endonuclease that cleaves target DNA with sequence specificity. There are multiple kinds of CRISPR/Cas systems based on the type of nuclease employed, Cas9 being the type II bacterial CRISPR/Cas system and most commonly engaged in CRISPR/Cas technique. CRISPRs are present only in the native bacterial system and are not a part of the CRISPR/Cas technology that is being used worldwide, as these sequences occur as a memory system in bacteria. CRISPRs are derived from plasmid or viral sequences, which are cut by Cas and integrated into the bacterial genome. Upon infection by a virus or any plasmid, CRISPRs are transcribed into CRISPR RNAs (CrRNA), which are complementary to the viral or plasmid genome, thereby marking the foreign DNA for cleavage by the Cas protein upon subsequent infections by similar viruses. crRNAs act in conjugation with trans-activating crRNAs (tracrRNAs). In CRISPR/Cas technique, crRNAs together with tracrRNA, are designated as complementary target genome known as single guide RNAs (sgRNAs). Also, the Cas protein does not undertake self-cleavage owing to the absence of protospacer-adjacent motif (PAM) in the bacterial genome (Jinek et al. 2012). Cas activates the mechanism for DNA repair by inserting DSBs into target DNA which are restore mostly with the inherent error-prone Non-homologous End Joining (NHEJ) mechanism producing genomic rearrangements, indels (gene knock-outs) and rarely Homology-Directed Repair (HDR) to generate insertions (gene knock-ins) (Fig. 8.1 and Table 8.1).

8.2 CRISPR/Cas-Based Genome Editing Tools

Genome editing using CRISPR/Cas has been progressively used as a technique for the development of new traits in crops through targeted manipulation in the target gene. The advancement of the GE-CRISPR/Cas, in particular, has enabled widespread use of genome editing and this use has greatly changed different areas of plant biology. sgRNA architecture is one of the principal factors in effectively editing target genes using CRISPR/Cas9. There are various platforms/online tools available to design sgRNA and for productive and unique target motifs selected in plants (Table 8.2).

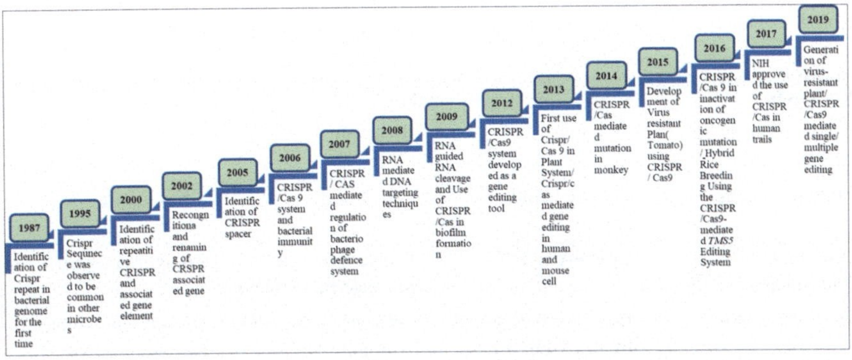

Fig. 8.1 An overview of major milestones in the development of CRISPR/Cas9 based genome editing approach

8.3 CRISPR/Cas9 System for Monocot and Dicot Plants

CRISPR/Cas is a new transgenic free editing tool, was first identified in the Gram-positive bacteria named *Streptococcus pyogenes* of adaptive immune system to bring about targeted changes in the genome. More than 20 crop plants, both monocots and dicots have been reported so far to adopt gene editing by the method of CRISPR/Cas9 for improving stress tolerance and yield management (Ricroch et al. 2017). Rice, wheat and maize are some prime examples of monocot plants where CRISPR/Cas has been successfully used for editing. Cereals are imperative part of our diet providing major health benefits with extensive nutritional value, making cereals the back bone of food security worldwide. Therefore, improving cereals is important and therefore, CRISPR/Cas has been applied widely to produce superior varities with enhanced production (Ansari et al. 2020). Rice genome has a plenty of potential PAMs of about 1 in every 10 base pair (bp) (Xie and Yang 2013). Shan et al. (2013) were amongst the first ones to report sequence-specific genome editing of three genes used in rice viz. phytoenedesaturase (*OsPDS*), betaine aldehyde dehydrogenase (*OsBADH2*) and mitogen-activated protein kinase (*OsMPK2*) by CRISPR/Cas9. These genes take part in checking out the reactions to diverse abiotic stress stimuli in crop plants and were edited for the first time, involving the use of protoplast transfection and particle bombarded in rice calli systems. For phytoenedesaturase (*OsPDS*) and betaine aldehyde dehydrogenase (*OsBADH2*) genes editing rates were observed at nine and seven percent. Mitogen-activated protein kinase (*OsMPK2*) gene is the negative stress regulator in rice. It was selected for targeted mutagenesis involving the use of three gRNAs in rice protoplasts, each gRNA complementary to 20 bp sequences in *OsMPK2* gene. Number of genes, including *OsDERF1*, *OsPMS*, *OsEPSPS*, *OsMSH1*, *OsMYB5*, have been evaluated in mutant rice lines where the CRISPR/Cas9 efficiency induces the aimed mutation (Zhang et al. 2014). Liu et al. (2012), recorded positive targeted ethylene response factor (*ERF*) editing in rice to increase blast resistance following

Table 8.1 Descriptive comparison among three major genome editing techniques

	CRISPR/Cas	TALEN	ZFN
Nucleases	Cas9	Fok1	Fok1
Recognition/target length	20 bp	30–40 bp	18–36 bp
Binding specificity	1:1 Nucleotide pairing	1Nucleotide	3 Nucleotide
DNA binding determinants	Cr RNA/sg RNA	Transcription activator-like effector	Zn figure protein
Mode of Action	DNA break targeted by RNA-DNA base complementarity	DNA break targeted by protein-DNA recognition	DNA break targeted by protein-DNA recognition
Targeting efficiency	High with most gRNAs	High with most nucleases	Variable
Nuclease design	Simple and most easy	complex but Feasible	Difficult and more complex
Modification efficiency	High	High	Low
Mutation rate %	20	20	10
Off-target effect	High potential for off-target (varies with conditions)	Low	High
Requirement for target site	Targeted sequence must precede a PAM	5′ target base must be preceded by T	Difficult to target G rich sequences
Methylation sensitivity	No	Yes	Yes
Multiplexable	Yes	Rare	Rare
Efficiency of multiple modifications	High	Low	Low
Cytotoxicity	Low	Low	Variable
Specificity	High	High	Low
Target delivery	Easy	medium to difficult	medium to difficult

Magnaporthe oryzae. There are many more cases of editing in rice, making rice most frequently targeted crop plants for genome editing via CRISPR/Cas.

Wheat forms an important staple food crop, grown worldwide; therefore, a great deal of work is being put into the utilization of CRISPR/Cas9 in generating targeted modifications in the wheat genome. But its complex hexaploid genome makes it very tough to target all the homologs at the same time to achieve requisite phenotypic characteristics. Still multiple cases of successful gene editing have been demonstrated in this tough crop. A null edits of *TaMLO* gene were generated in wheat thereby conferring tolerance to powdery mildew disease caused by *Blumeria graminis f. sp. tritici* (Wang et al. 2014a, b). Two genes related to stress particularly abiotic stress in wheat protoplast, namely *TaDREB2* (wheat *dehydration responsive element binding protein* 2) and *TaERF3* (wheat *ethylene-responsive factor 3*) were successfully edited

Table 8.2 Various types of CRISPR/Cas tools available online for designing sgRNA and targeting motifs in plants

Tool	Function	Type of CRISPR/Cas system	Input	Off-target analysis	Source	Reference
CRISPR Multitargeter	CRISPR guide RNA targets	Multiple types	Sequence/identifiers	No	https://www.multicrispr.net/	Prykhozhij et al. (2015)
CRISPR-ERA	Cas9 target design tool for genome editing, repression and activation	Type II	Sequence/identifiers	Yes	https://crispr-era.stanford.edu/	Liu et al. (2015)
sgRNA Designer	Guide RNA design tool	Type II	Sequence/identifiers	No	https://portals.broadinstitute.org/gpp/public/analysis-tools/sgrna-design	Doench et al. (2014a, b)
CRISPRscan	Design Cas9/Cas 12a target	Type II	Sequence/identifiers	Yes	https://www.crisprscan.org/	Moreno-Mateos et al. (2015)
SSC	sgRNA design	Type II	Sequence/identifiers	No	https://cistrome.org/SSC/	Xu et al. (2011)
sgRNA scorer	Design sgRNA	Type II	Sequence/identifiers	Yes	https://omictools.com/sgrna-scorer-tool	Chari et al. (2015)
WU-CRISPR	Design sgRNA	Type II	Sequence/identifiers	Yes	https://crispr.wustl.edu/	Wong et al. (2015)
E-CRISP	Target sites design tool	Different Type II	Sequence/identifiers	Yes	https://www.e-crisp.org/E-CRISP/	Heigwer et al. (2014)

(continued)

Table 8.2 (continued)

Tool	Function	Type of CRISPR/Cas system	Input	Off-target analysis	Source	Reference
CRISPR-Design	Design and analysis of guide RNA	Type II	Sequence only	Yes	https://dharmacon.horizondiscovery.com/gene-editing/crispr-cas9/crispr-design-tool/	Hsu et al. (2013)
CRISPRdirect	Design sgRNA with reduced off-target site	Type II	Sequence/identifiers	Yes	https://crispr.dbcls.jp/	Naito et al. (2014)
Cas9-Design	sgRNA design tool	Type II	Sequence only	Yes	https://omictools.com/cas9-design-tool	Ma et al. (2013)
Zi-Fit-Targeter	Gene editing and expression, identify ZFN sites	Type II	Sequence only	No	https://zifit.partners.org/ZiFiT/	Hwang et al. (2013)
CRISPR-P	sgRNA design in plant	Type II	Sequence only	Yes	https://crispr.hzau.edu.cn/CRISPR2/	Lie et al. (2014a, b)
CHOPCHOP	Target site binding tool	Different Type II	Sequence/identifiers	Yes	https://chopchop.cbu.uib.no/	Montague et al. (2014)
CC-Top	Target prediction tool for Cas9/Cas12a	Type II	Sequence only	Yes	https://crispr.cos.uni-heidelberg.de/	Stemmer et al. (2015)

(continued)

Table 8.2 (continued)

Tool	Function	Type of CRISPR/Cas system	Input	Off-target analysis	Source	Reference
Cas online designer	Design Cas9 with its variants and Cas 12a targets	Type II	Sequence only	Yes	https://www.rgenome.net/cas-designer/	Guo et al. (2015)
sgRNAcas9	sgRNA design & potential off-target sites prediction tool	Type II	Sequence only	Yes	www.biootools.com	Xie et al. (2014)
CRISPRseek	Target specific guide RNA design tool	Different Type II	Sequence only	Yes	https://www.bioconductor.org/packages/release/bioc/html/CRISPRseek.html	Zhu et al. (2014)
CAS-OFFinder	Cas9 RNA-guided endonucleases (potential off-target sites)	Type II	crRNA sequences	Yes	https://www.rgenome.net/cas-offinder/	Bae et al. (2014)
GT-Scan	Find potential off-targets	Type II	Sequence only	Yes	https://gt-scan.csiro.au/	O'Brien and Bailey (2014)

(continued)

Table 8.2 (continued)

Tool	Function	Type of CRISPR/Cas system	Input	Off-target analysis	Source	Reference
Cas-OT	Genome-wide Cas9/gRNA off-target searching tool	Type II	Sequence/identifiers	Yes	https://omictools.com/casot-tool	Xiao et al. (2014)
Addgene	For genome engineering, transcriptional regulation, RNA targeting	Type II	Sequence/identifiers	Yes	https://www.addgene.org/crispr/	Kamens (2015)
CRISPR genome analyzer	Genome editing experiment analysis platform	Type II	Sequence only	Yes	https://crispr-ga.net/	Güell et al. (2014)
RGEN Tool	Potential off-target sites prediction tool	Type II	Sequence/identifiers	Yes	https://www.rgenome.net/	Cho et al. (2013)
AGESeq	Analysis of genome editing by sequencing	Type II	Sequence only	Yes	https://aspendb.uga.edu:8085/	Xue and Tasi (2015)

by CRISPR/Cas9 (Kim et al. 2018). CRISPR/Cas9 has also been applied to *Zea mays* effectively. Phytic acid, an anti-nutritional substance and environmental pollutant, makes about 70% of the maize seed. CRISPR/Cas9 mediated knockout of genes namely, *ZmIPK*, *ZmIPK1A*, and *ZmMRP4* which are implicated in the synthesis of phytic acid in *Z. mays* has also been accomplished, making the maize plants more nutritious and phytic acid-free. For improving essential characters in other monocot crops, CRISPR/Cas9 genome editing approach has proved very efficient. Successful knockout of endo-N-acetylb-D-glucosaminidase (*ENGase*) gene in barley by CRISPR/Cas9 system has been reported by Kapusi et al. (2017). There is a stable expression and successful integration of CRISPR/Cas9 in the host genome.

In dicots, genome editing mediated by CRISPR/Cas9 was first reported in *Arabidopsis* by Feng et al. (2013). This system edited genes in *Arabidopsis* mainly related to albinism like *magnesium-chelatase subunit I* (*CHLI1*) and *CHLI2*. CRISPR/Cas9 technique was also used to edit *Arabidopsis TRANSPARENT TESTA4 (TT4)* gene. Pyott et al. (2016a, b) demonstrated that sequence-specific point mutations at eukaryotic translation initiation factor locus *eIF(iso)4E* in *Arabdopsis* results in engineering the absolute tolerance to Turnip mosaic virus *(TuMV)*. Cotton forms an important fiber crop among dicots, therefore, targeting cotton has been major concern in several labs. CRISPR/Cas9 technology has widely used for producing mutations in cotton homologous genes (Gao et al. 2017). Tomatoes form an ideal crop for dicots to examine the CRISPR/Cas9-based gene editing due to the existence of an effective method of transformation. By using editing systems, generation of parthenocarpic tomato plants could be achieved by making site-specific somatic mutations mainly in *SlIAA9* related parthenocarpy (Ueta et al. 2017). This editing system has also been used in developing viral resistance in cucumber as well. Targeted genome editing of the gene *4E (eIF4E)* provided tolerance to various viral pathogens. To increase the shelf life of fruits and improvement of horticultural plants as well as vegetables CRISPR/Cas9 editing system is being vastly utilized (Fig. 8.2).

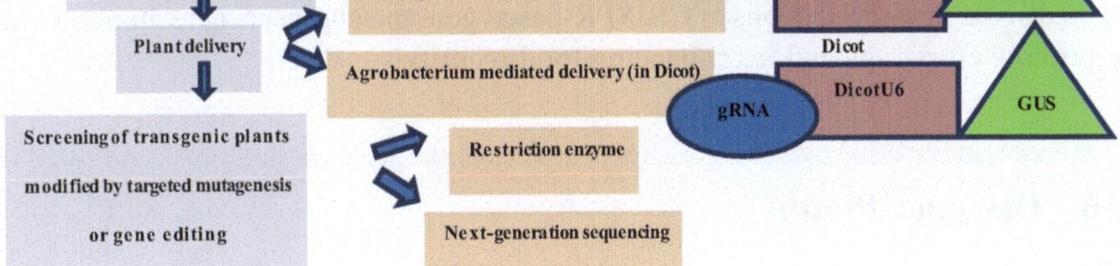

Fig. 8.2 Overview of generalized method for CRISPR/Cas9 based gene editing in crops

8.4 Application of CRISPR/Cas9 Systems in Plant Genomic Research

The genome editing space has seen unparalleled scientific alteration in genomic research. Due to advancement in high-throughput sequencing and improvements in genome engineering tools exact location and structure of functional element with their manipulation can be easily studied to control any genetic material. In the coming decades, genome editing will take part in an improved chunk in crop development efforts to achieve food security with full potential by overcoming its limitations. The CRISPR/Cas9 is a valuable technology for functional validation of genes through either knock-in or knock-out as well as chromatin alteration of the targeted gene loci in different types of cells and organisms. It can be applied in various areas including molecular breeding research and genetic improvement, secondary metabolism and synthetic biology research, multiplex genome targeting, growth and development, reverse genetics, miRNA function and regulation, and biotic/abiotic resilience and so on (Vats et al. 2019). The current chapter briefly introduces about the most recent development in CRISPR/Cas gene rephrasing system and their perspective function in translational research in plants (Fig. 8.3).

8.5 Applicability and Concerns of CRISPR/Cas-Based Genome Editing for Plant Breeding and Genetic Improvement

CRISPR/Cas9 has been extensively used in crop improvement to include novel plant traits into a variety of crops so far to generating dwarf and semi-dwarf plants, regulating flowering time and fruit maturation in plants, enhancement of seed weight and number and switching of cytoplasmic male sterile line into fertile lines, which can be achieved by double-strand break (DSB) induction, resulting in mutations by non-homologous end joining (NHEJ). CRISPR/Cas9 genome editing technology leads to an innovative outcome on plant breeding research (Weeks et al. 2015; Belhaj et al. 2015; Ma and Liu 2016). The recognition of target alleles that display unique phenotypic traits provides an excellent opportunity to generate transgenic free prompt and effective improvable cultivars using genome editing approaches. Here, we have highlighted some key applications of CRISPR-based genome engineering technology for improving crop traits and for enhancing food security.

8.6 Designer Plants

Designer plants were produced instead of a total loss of function by functional modifications of *BolC.GA4a* null edits in rapeseed; *ER1*, *ER2* and *SEC3a* genes in *Oryzae*

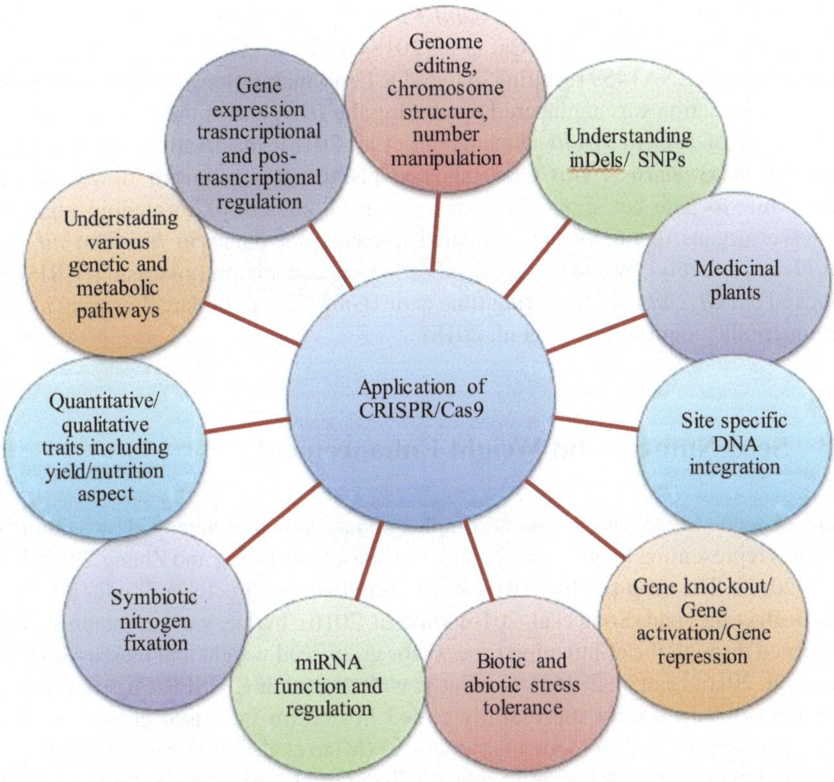

Fig. 8.3 List of most significant applications of CRISPR/Cas9 based genome editing in plants

sativa and *Dep1* in wheat resulted in dwarf and semi-dwarf phenotypes (Lawrenson et al. 2015; Zhang et al. 2016, 2018; Ma et al. 2018). Knock-down in the *BaA6.RGA* gene has also resulted in semi-dwarf rapeseed mutants (Yang et al. 2017). Targeted knock-out of CCD7 that regulates a significant part in SL biogenesis course in rice using CRISPR/Cas resulted in mutant with increased tillering and reduced height (Butt et al. 2018).

8.7 Regulation of Flowering Time and Fruit Maturation in Plants

The circadian rhythm of plant developmental process from the vegetative to reproductive phase is also an important determinant of yield. The engineering of cultivars based on these traits should be considered with respect to various crop husbandry and habitats. Null edits of *OsHd2, Hd4, Hd5* and *SlSP5G* genes, resulted in increased maturity in rice and tomato, while *GmFT2* in soybean altered floral initiation (Li et al. 2017b; Soyk et al. 2017; Cai et al. 2018). Li et al. (2018) reported the alteration in tomato RNA1459 resulting in overdue berry maturation. *Setaria viridis* with late floral initiation was engineered by altering the *ID1* gene of the *Zea mays* which is homolog of *Setaria viridis* (Jaganathan et al. 2018). Inactivation of *carotenoid cleavage dioxygenase 4* (*InCCD4*) showed phenotypic change of flower colour from white to pale but significantly increased the content of carotenoid pigment in flower organs (petal) of the generated mutant *ccd4* plants in *Ipomoea nil*, cv. AK77 respectively (Watanabe et al. 2018). Soybean plants possessing CRISPR knocked out *Glycine max* flowering time gene (*GmFT2*) exhibit late flowering under photoperiodic conditions (Cai et al. 2018).

8.8 Seed Number and Weight Enhancement

Grain or seed number and weight is complex computable trait managed by a number of genes representing many more direct yield causations (Xing and Zhang 2010; Bai et al. 2012). The crop productivity can be increased by directly null edits of anti-genes affecting yield (Song et al. 2016; Ma et al. 2016). In rice, various mutants were generated through the elimination process for grain/seed weight and increased yield (Xu et al. 2016; Li et al. 2016a). Similarly, with the aid of CRISPR/Cas9 approach, a number of genes were successfully edited in rice for increased crop yield and alterations in various developmental processes (Miao et al. 2018). Similar work was done by targeting *GASR7* gene in wheat (Zhang et al. 2016). Knock-out of *CLV3* gene homoalleles increased seed count and silique as well as locule type in model crop tomato and Brassica (Yang et al. 2018; Rodríguez-Leal et al. 2017). Seedless fruits were obtained in tomato by Cas9-mediated knock-out of *AGL6* (Klap et al. 2017). In addition, null edits of the rapeseed gene namely, *ALCATRAZ* resulted in reduced seed shattering (Braatz et al. 2017).

8.9 Cytoplasmic Male Sterility (CMS) Shift

In hybrid breeding sense, an important prerequisite is the possibility of converting between male fertility and sterility. Interestingly, an array of genes related to pollen development were successfully knocked out in rice and maize for producing male sterile crops (Svitashev et al. 2015, 2016; Zhou et al. 2016; Li et al. 2016c; Zou et al. 2018). Another key gene related to male sterility in rice was also successfully edited by Shen et al. 2017. Additionally, haploid technology is generally considered as the most key technology in crop breeding programmes. Recently, *MALT* gene nulls were used to produce haploid activator accessions in rice. Furthermore, these null *MALT* accessions were further tested as pollinators resulting in a low number of haploids (Yao et al. 2018). In Maize plants, gene editing of *ZmTMS5* gene related to male sterility were found to stimulate male sterility and may have value in hybrid seed production (Li et al. 2017b).

8.10 Nutritional Value and Quality Trait Improvement in Plants

Various tools have been used to improve the nutritional quality of food and feed, site-directed mutagenesis mediated modification is one of them. IPK gene plays a very important role in phytic acid biosynthesis, and knockdown of this gene causes phytic acid reduction in maize (Liang et al. 2014). Sun et al. 2017 studies highlighted that editing of two key genes in rice namely *SBE1* and *SBEIIb* significantly decreases the content of amylopectin which could decrease the chances of developing diabetes II. In another study by Shan et al. (2015) in rice using TALEN technology betaine aldehyde dehydrogenase 2 (*OsBADH2*) gene, a key contributor to fragrance was targeted to alter the expression. Besides rice, the gene editing study is also being performed in wheat. Sánchez-León et al. (2018) concurrently edited an array of genes in bread wheat which not only decreases the total gluten content but also were found to have weak immunoreactivity. In *Camelina sativa*, a key oil-producing gene *FAD2* was successfully knocked out to increase the content of monounsaturated oleic acid (Jiang et al. 2017; Morineau et al. 201). Recently, CRISPR/Cas tool was employed in the same crop where *FAE1* gene was knocked out to change the fatty acids profile (Ozseyhan et al. 2018). Besides, another gene was Wx1 was successfully knocked out in maize plants (Waltz, 2016). In *Agaricus bisporus*, *polyphenol oxidase* (*PPO*) gene was edited for anti-browning properties. Previous studies have revealed that knockout of vacuolar invertase genes by TALEN produces potatoes without acrylamide (Clasen et al. 2016). Alternatively, editing of potato *granule-bound starch synthase (GBSS)* gene leads to the production of no amylose and at the same time amylopectin-rich starch in potato plants (Andersson et al. 2017).

8.11 CRISPR/Cas9 in Biotic Stress Tolerance

Plants are constantly confronted by a variety of phytopathogens which reduce overall crop yield and quality thereby causing huge economic losses (Taylor et al. 2004; Savary et al. 2012). Therefore, producing plants resistant to biotic stress has been one of the foremost priorities of researchers. Site-directed mutagenesis has been applied to develop bacterial resistance crops using RNA-guided Cas endonucleases. Knock-out of two genes namely *SWEET13* and *ERF922* in rice successfully provides disease resistance to blast disease-causing bacterial pathogens (Zhou et al. 2015;

Wang et al. 2016a, b). For developing citrus canker disease resistance, editing of the promoter sequence, and knocking out of the *CsLOB1* gene was carried out (Jia et al. 2016 and Peng et al. 2017). Recently, gene knockout of *WRKY52* in grape wine decreases immunity to *Botrytis cinerea,* a fungal pathogen (Wang et al. 2018). Previous reports have shown that inactivating of *MLOI* in tomato and wheat as well as *EDR1* gene in wheat, leads to powdery mildew disease resistance (Wang et al. 2014a, b, Nekrasov et al. 2017; Zhang et al. 2017). Interestingly, editing of *Theobroma cacao* defense suppressor gene viz, *NON- EXPRESSOR OF PATHOGENESIS-RELATED GENES 3* (*NPR3*) improves immunity not only to fungal pathogens but also priming the immune response by increasing the expression of a wide range defense signature genes (Fister et al. 2018). Targeted mutation of 4E(*eIF4E*) recessive gene was found to be tolerant against various plant viral pathogens. Li et al. (2019) reported that CRISPR/Cas9-Mediated mutagenesis (*SlNPR1*) decreases drought tolerance in the tomato plant. Ectopic expression of *AtNPR1* amplifies disease tolerance in a range of crops including model crop *Arabidopsis,* as well as various vegetables and fruit crops (Cao et al. 1998; Wally et al. 2009; Dutt et al. 2015; Malnoy et al. 2007; Le et al. 2011).

8.12 Applications of CRISPR/Cas9-Based Genome Editing for Abiotic Stress Tolerance

Abiotic stresses such as waterlogging, salinity, heat, drought significantly reduces total yield and productivity in most of the agriculturally important crops (see Pandey et al. 2017). CRISPR/Cas9 for drought stress tolerance in maize was developed recently by DuPont scientists by modifying the anti-switch of ethylene-responsive gene (*ARGOS8*) (Shi et al. 2017). Previous reports have documented that the editing of *ethylene-responsive factor 3* and *dehydration responsive element binding protein 2* in wheat respectively (Kim et al. 2018) provides abiotic tolerance. In rice, salt tolerance was studied using Cas9-induced knockout lines *(NCED3)* (Huang et al. 2018). Drought tolerance has been enhanced by knockout of *MAPK3* as well as an increased transcription of *ARGOS8* gene in tomato and corn, separately (Wang et al. 2017a; Shi et al. 2017). In soybean, knock-outs of *drb2a* and *drb2b* genes lead to not only drought but also salt tolerance. Similarly, seed germination under heat stress increases in Cobham green and lettuces by knockout of the *9-cis*-Epoxycarotenoid Dioxygenase 4 *(NCED4)* gene. Table 8.3 displays the list of traits edited by genome editing technology.

Table 8.3 An overview of genome editing in various crops from the perspective of application and most commonly targeted traits

Application perspective	Crop	Genome editing method/strategy	Target gene	Targeted trait	Reference
Yield and quality					
		CRISPR/Cas 9	*TGW6, GW2* and *GW5*	Increases grain weight	Xu et al. (2016a)
	Rice	–	Hd2, Hd4, and Hd5	For early maturity	Li et al. (2017a, b)
			CSA	Photoperiod modulated MS lines	Li et al. (2016a, d)
		CRISPR/Cas 9	Gn1a, DEP1, GS3, and IPA1	Increase of grain number, panicle and grain size, and plant shape and size	Li et al. (2016a)
			BADH2	Enriched aroma	Shao et al. (2017)
			PYLs	Plant architecture and yield	Miao et al. (2018)
	Wheat	CRISPR-Cas9	TaGW2	Grain type	
	Wheat		sd1	Discovering number of high-yield genes	Huang et al. (2018)

(continued)

Table 8.3 (continued)

Application perspective	Crop	Genome editing method/strategy	Target gene	Targeted trait	Reference
	Sorghum	CRISPR-Cas9	Alpha-Kafirin Gene Family	PDCAAS	Li et al. (2018)
Biotic stress tolerance					
		TALENs	OsSWEET13	Enhanced resistance to bacterial blight	Blanvillain-Baufum et al. (2017)
		TALENs	Os09g29100	Bacterial leaf streak tolerance	Cai et al. (2017)
	Rice	CRISPR/Cas9	OsERF922	Blast resistance	Wang et al. (2016a, b)
	A. thaliana	NHEJ	eIF(iso)4E	Resistance to Turnip mosaic virus (TuMV)	Pyott et al. (2016)
	Cucumber	NHEJ	eIF4E	Poty-virus resistance	Chandrasekaran et al. (2016)
Abiotic stress tolerance					
		CRISPR/Cas 9	OsEPSPS	Glyphosate resistant	Li et al. (2016c)
	A. thaliana	HDR	MIR169a	DT	Zhao et al. (2016)
	A. thaliana	NHEJ	OST2	Stomata closure response and ABA	Osakabe et al. (2016)
		TALENs Li	OsALS	Herbicide resistant	Li et al. (2016b)
		CRISPR/Cas 9	ALS	Herbicide resistant	Sun et al. (2016a, b)
		Base editing	C287	Herbicide resistant	Shimatani et al. (2017)
		CRISPR/Cas 9	Os SAPK2	Drought resistance	Lou et al. (2017)

(continued)

Table 8.3 (continued)

Application perspective	Crop	Genome editing method/strategy	Target gene	Targeted trait	Reference
	Rice	NHEJ	OsHAK-1	Low cesium accumulation	Cordones et al. (2017)
	Maize	HDR	ARGOS8	Yield and DT	Shi et al. (2017)
	Tomato	NHEJ	SlMAPK3	Drought tolerance	Wang et al. (2017a, b)
	Tomato-	CRISPR-Cas9	SlCBF1	Plant chilling tolerance	
	Rice	NHEJ	OsPRX2	Potassium deficiency resistance	Mao et al. (2018)
	Rice	CRISPR/Cas 9	TMS5	Thermo-sensitive genic male sterility (TGMS)	Barman et al. (2019)
	Tomato	CRISPR/Cas9	SlNPR1	DT	Li et al. (2019)
	Potato	CRISPR/Cas9	Coilin	Virus resistance and stress tolerance	Makhotenko et al. (2019)
Nutritional improvement					
	Potato	HDR	ALS1	Herbicide resistance	Butler et al. (2016)
	Soybean	NHEJ	GmPDS11 and GmPDS18	Carotenoid biosynthesis	Du et al. (2016)
	Rice	CRISPR/Cas 9	OsNRAMP5	Low cadmium content	Tang et al. (2017)
	Rice	CRISPR/Cas 9	SBEIIb and SBEI	Production of high amylose rice	Sun et al. (2017)

(continued)

Table 8.3 (continued)

Application perspective	Crop	Genome editing method/strategy	Target gene	Targeted trait	Reference
		BE	PDS, SBEIIb	Nutritional enhancement	Li et al. (2017c)
	Cassava	NHEJ	MePDS	Biosynthesis of carotenoid	Odipio et al. (2017)
	Wheat	HDR	TaVIT2	Iron content	Connorton et al. (2017)

8.13 Utility of CRISPR/Cas-Based Genome Editing in MiRNA Function and Regulation

MicroRNAs (miRNAs) are a group of aspecial type of RNAs which are generally non-coding in nature and identify the target mRNAs through a unique base-pairing, thereby controlling an array of cellular, developmental and stress responsive processes (Chen 2012; Kumar 2014). In plants, CRISPR/Cas9 has recently been recognized as a key technique to produce miRNA knockout mutants, that are generally finer for reverse genetics. Interestingly, in Zebrafish miRNA gene clusters have been edited through multiple sgRNAs (Narayanan et al. 2016). With multiple gene targeting of CRISPR/Cas9 systems (Lowder et al. 2015), it could be widely used in plants to generate multiple simultaneous miRNA gene family edits for genetic studies. Recently, in model plant Arabidopsis and legume crop soybean two miRNA genes (Gm*miR1514*, and *miR1509)* and *A. thaliana miR169a*, and *miR827a*) were edited via the CRISPR/Cas9 system (Jacobs et al. 2015; Zhao et al. 2016).

8.14 Applications of CRISPR/Cas9-Based Genome Editing for Increasing Herbicide Resistance in Plants

Herbicide resistance is one of the areas of great potential interest. In nature, crop weed competition is intense severely affecting the crop yield. Labor availability is always a problem, more so in developed nations. Some of the target genes providing herbicide tolerance have been targeted (see Lombardo et al. 2016). ZFN was used for the first time, to induce targeted mutation in the *ALS* gene in the tobacco followed by several commercial crops using TALENs and CRISPR/Cas9 tools (Cai et al. 2009; Shukla et al. 2009; Townsend et al. 2009; Butler et al. 2015; Svitashev et al. 2015; Li et al. 2016; Sun et al. 2016a, b). In addition, mutation in the *EPSPS* gene in flax and rice using CRISPR/Cas9 resulted in glyphosate-tolerant crops (Sauer et al. 2016, Li et al. 2016c).

8.15 Conclusions and Future Perspective

In crop biotechnology, CRISPR/Cas9 has emerged as one of the most promising technology for increasing crop yield and improving quality as well as tolerance to multiple stresses. It has unique features including precise and easy manipulation, and high efficiency, and requires much less efforts in terms of getting ethical clearance, and hence can be widely used in the field of applied research. In general, induced mutations mediated by CRISPR/Cas9 are largely similar to the mutations generated by random mutagenesis. Furthermore, the marker and Cas9 genes can be easily removed through segregation in transgenic plants, thereby developing transgenic free plants which is the key feature of genome editing making the regulations around edited plants less stringent. Despite its wide applications in crop biotechnology, there are few challenges like optimization of the function of Cas9 as well as minimizing off-target rates. The primary concerns about the efficient exploration of technological advances in genome editing approaches include regulatory procedures not yet formulated in most countries; many crop species are challenging to transform; and DNA free methods are limited to few laboratories. The acceptance of genome edited plants with single base change or mutation mimicking natural variation looks easy; however, knock-in of the entire gene or more extensive manipulations seems to follow exhaustive regulatory guidelines. Similarly, only a handful of crop species have a well-standardized genetic transformation protocol; therefore, adopting genome editing for crops where transformation protocols are not optimized looks complicated. Compared to DNA transformation-based methods, DNA free genome editing approaches have less concern about acceptability. Even though adopting DNA free method is challenging due to the low frequency for getting desired manipulations, costly, and involved highly sophisticated techniques. However, the present boom in the genome editing field will surely deliver several advances to overcome these limitations and hopefully help to bring a second green revolution.

References

Andersson M, Turesson H, Nicolia A, Fält A, Samuelsson M, Hofvander P (2017) Efficient targeted multiallelic mutagenesis in tetraploid potato (*Solanum tuberosum*) by transient CRISPR-Cas9 expression in protoplasts. Plant Cell Rep 36:117–128

Ansari WA, Chandanshive SU, Bhatt V, Nadaf AB, Vats S, Katara JL, Sonah H, Deshmukh R (2020) Genome editing in cereals: approaches, applications and challenges. Int J Mol 21(11):4040

Bae S, Park J, Kim JS (2014) Cas-OFFinder: a fast and versatile algorithm that searches for potential off-target sites of Cas9 RNA-guided endonucleases. Bioinformatics 30:1473–1475

Bai X, Bi W, Yongzhong X (2012) Yield-related QTLS and their applications in rice genetic improvement. J Integr Plant Biol 54:300–311

Barman HN, Sheng Z, Fiaz S, Zhong M, Wu Y, Cai Y, Hu P (2019). Generation of a new thermo-sensitive genic male sterile rice line by targeted mutagenesis of TMS5 gene through CRISPR/Cas9 system. BMC Plant Biol 19(1):109

Belhaj K, Chaparro-Garcia A, Kamoun S, Patron NJ, Nekrasov V (2015) Editing plant genomes with CRISPR/Cas9. Curr Opin Biotechnol 32:76–84

Blanvillain-Baufum S, Reschke M, Sol M, Auguy F, Doucoure H, Szurek B et al (2017) Targeted promoter editing for rice resistance to *Xanthomonas oryzae* pv. oryzae reveals differential activities for SWEET14-inducing TAL effectors. Plant Biotechnol J 15:306–317

Braatz J, Harloff H, Mascher M, Stein N, Himmelbach A, Jung C (2017) CRISPR-Cas9 targeted mutagenesis leads to simultaneous modification of different homoeologous gene copies in polyploid oilseed rape (*Brassica napus*). Plant Physiol 174:935–942

Butler NM, Atkins PA, Voytas DF, and Douches DS (2015) Generation and inheritance of targeted mutations in potato (*Solanum tuberosum* L.) using the CRISPR/Cas system. PLoS One 10:e0144591

Butler NM, Baltes NJ, Voytas DF, and Douches DS (2016) Geminivirus-mediated genome editing in potato (*Solanum tuberosum* L.) using sequence-specific nucleases. Front Plant Sci 7:1045

Butt H, Jamil M, Wang JY, Al-Babili S, Mahfouz M (2018) Engineering plant architecture via CRISPR/Cas9-mediated alteration of strigolactone biosynthesis. BMC Plant Biol 18:174

Cai CQ, Doyon Y, Ainley WM, Miller JC, Dekelver RC, Moehle EA et al (2009) Targeted transgene integration in plant cells using designed zinc finger nucleases. Plant Mol Biol 69:699–709

Cai L, Cao Y, Xu Z, Ma W, Zakria M, Zou L et al (2017) A Transcription activator-like effector Tal7 of *Xanthomonas oryzae* pv. oryzicola activates rice gene Os09g29100 to suppress rice immunity. Sci Rep 7:5089

Cai Y, Chen L, Liu X, Guo C, Sun S, Wu C, Jiang B, Han T, Hou W (2018) CRISPR/Cas9-mediated targeted mutagenesis of GmFT2a delays flowering time in soya bean. Plant Biotechnol J 16:176–185

Cao H, Li X, Dong X (1998) Generation of broad-spectrum disease resistance by overexpression of an essential regulatory gene in systemic acquired resistance. Proc Natl Acad Sci USA 95(11):6531–6536

Chandrasekaran J, Brumin M, Wolf D, Leibman D, Klap C, Pearlsman M et al (2016) Development of broad virus resistance in non-transgenic cucumber using CRISPR/Cas9 technology. Mol Plant Pathol 17:1140–1153

Chari R, Mali P, Moosburner M, Church GM (2015) Unraveling CRISPR-Cas9 genome engineering parameters via a library-on-library approach. Nat Methods 12:1–7

Chen X (2012) Small RNAs in development—insights from plants. Curr Opin Genet Dev 22:361–367

Cho SW, Kim S, Kim JM, Kim JS (2013) Targeted genome engineering in human cells with the Cas9 RNA-guided endonuclease. Nat Biotech 31(3):230–232

Clasen BM, Stoddard TJ, Luo S, Demorest ZL, Li J, Cedrone F, Tibebu R, Davison S, Ray EE, Daulhac A, Coffman A, Yabandith A, Retterath A, Haun W, Baltes NJ, Mathis L, Voytas DF, Zhang F (2016) Improving cold storage and processing traits in potato through targeted gene knockout. Plant Biotechnol J 14(1):169–176

Connorton JM, Jones ER, Rodriguez-Ramiro I, Fairweather-Tait S, Uauy C, Balk J (2017) Wheat vacuolar iron transporter TaVIT2 transports Fe and Mn and is effective for biofortification. Plant Physiol 174:2434–2444

Cordones MN, Mohamed S, Tanoi K, Natsuko Kobayashi NI, Takagi K, Vernet A et al (2017) Production of low-Cs+rice plants by inactivation of the K+ transporter OsHAK1 with the CRISPR-Cas system. Plant J 92:43–56

Doench JG, Hartenian E, Graham DB, Tothova Z, Hegde M, Smith I, Sullender M, Ebert BL, Xavier RJ, Root DE (2014a) Rational design of highly active sgRNAs for CRISPRCas9-mediated gene inactivation. Nat Biotechnol 32:1262–1267

Doench JG, Fusi N, Sullender M, Hegde M, Vaimberg EW, Donovan KF, Smith I, Tothova Z, Wilen C, Orchard R, Virgin HW, Listgarten J, Root DE (2014b) Optimized sgRNA design to maximize activity and minimize off-target effects of CRISPRCas9. Nat Biotechnol 34:184–191

Du H, Zeng X, Zhao M, Cui X, Wang Q, Yang H et al (2016) Efficient targeted mutagenesis in soybean by TALENs and CRISPR/Cas9. J Biotechnol 217:90–97

Dutt M, Barthe G, Irey M, Grosser J (2015) Transgenic citrus expressing an Arabidopsis NPR1 gene exhibit enhanced resistance against Huanglongbing (HLB; Citrus Greening). PLoS One 10(9):e0137134

Feng Z, Zhang B, Ding W, Liu X, Yang DL, Wei P et al (2013) Efficient genome editing in plants using a CRISPR/Cas system. Cell Res 23:1229–1232

Fister AS, Landherr L, Maximova SN, Guiltinan MJ (2018) Transient Expression of CRISPR/Cas9 Machinery Targeting TcNPR3 Enhances Defense Response in *Theobroma cacao*. Front Plant Sci 9(268):1–15

Gao W, Long L, Tian X, Xu F, Liu J, Singh PK et al (2017) Genome editing in cotton with the CRISPR/Cas9 system. Front Plant Sci 8:1364

Guo D, Li X, Zhu P, Feng Y, Yang J, Zheng Z (2015) Online high-throughput mutagenesis designer using scoring matrix of sequence-specific endonucleases. J Integr Bioinform 12:283

Güell M, Yang L, Church GM (2014) Genome editing assessment using CRISPR Genome Analyzer (CRISPR-GA). Bioinform 30(20):2968–2970

Heigwer F, Kerr G, Boutros M (2014) E-CRISPR: fast CRISPR target site identification. Nat Methods 11:122–123

Hsu PD, Scott DA, Weinstein JA, Ran FA, Konermann S, Agarwala V, Li Y, Fine EJ, Wu X, Shalem O, Cradick TJ, Marraffini LA, Bao G, Zhang F (2013) DNA targeting specificity of RNA-guided Cas9 nucleases. Nat Biotechnol 31:827–832

Huang Y, Guo Y, Liu Y, Zhang F, Wang Z, Wang H, Wang F, Li D, Mao D, Luan S, Liang M, Chen L (2018) 9-cis-Epoxycarotenoid dioxygenase 3 regulates plant growth and enhances multi-abiotic stress tolerance in rice. Front Plant Sci 9:162

Hwang WY, Fu Y, Reyon D, Maeder ML, Tsai SQ, Sander JD, Peterson RT, Yeh JJ, Joung JK (2013) Efficient genome editing in zebrafish using a CRISPR-Cas system. Nat Biotechnol 31:227–229

Jacobs TB, La Fayette PR, Schmitz RJ, Parrott WA (2015) Targeted genome modifications in soybean with CRISPR/Cas9. BMC Biotechnol 15:16

Jaganathan D, Ramasamy K, Sellamuthu G, Jayabalan S, Venkataraman G (2018) CRISPR for crop improvement: an update review. Front Plant Sci 9

Jia H, Zhang Y, Orboviæ V, Xu J, White Frank F, Jones Jeffrey B et al (2016) Genome editing of the disease susceptibility gene CSLOB1 in citrus confers resistance to citrus canker. Plant Biotechnol J 15:817–823

Jiang WZ, Henry IM, Lynagh PG, Comai L, Cahoon EB, Weeks DP (2017) Significant enhancement of fatty acid composition in seeds of the allohexaploid, *Camelina sativa*, using CRISPR/Cas9 gene editing. Plant Biotechnol J 15:648–657

Jinek M, Chylinski K, Fonfara I, Hauer M, Doudna JA, Charpentier E (2012) A programmable dual-RNA-guided DNA endonuclease in adaptive bacterial Immunity. Sci 337(6096):816–821

Kamens J (2015) The Addgene repository: an international nonprofit plasmid and data resource. Nucleic Acids Res 43(D1):D1152–D1157

Kapusi E, Corcuera-Gomez M, Melnik S, Stoger E (2017) Heritable genomic fragment deletions and small indels in the putative engase gene induced by CRISPR/Cas9 in barley. Front Plant Sci 8:540

Kim D, Alptekin B, Budak H (2018) CRISPR/Cas9 genome editing in wheat. Funct Integr Genomics 18:31–41

Klap C, Yeshayahou E, Bolger AM, Arazi T, Gupta SK, Shabtai S, Usadel B, Salts Y, Barg R (2017) Tomato facultative parthenocarpy results from SlAGAMOUS-LIKE 6 loss of function. Plant Biotechnol J 15:634–647

Klompe SE, Vo PLH, Halpin-Healy TS, Sternberg SH (2019) Transposon-encoded CRISPR–Cas systems direct RNA-guided DNA integration. Nature 571:219–225

Kumar R (2014) Role of microRNAs in biotic and abiotic stress responses in crop plants. Appl Biochem Biotechnol 174:93–115

Lawrenson T, Shorinola O, Stacey N, Li C, Østergaard L, Patron N, Uauy C, Harwood W (2015) Induction of targeted, heritable mutations in barley and *Brassica oleracea* using RNA-guided Cas9 nuclease. Genome Biol 16:258

Le HG, Farine S, Kieffer-Mazet F, Miclot AS, Heitz T, Mestre P, Bertsch C, Chong J (2011). *Vitis vinifera* VvNPR1.1 is the functional ortholog of AtNPR1 and its overexpression in grapevine triggers constitutive activation of PR genes and enhanced resistance to powdery mildew. Planta 234(2):405–417

Lei Y, Lu L, Liu HY, Li S, Xing F, Chen LL (2014a) CRISPR-P: a web tool for synthetic single-guide RNA design of CRISPR-system in plants. Mol Plant 7:1494–1496

Lei Y, Lu L, Liu HY, Li S, Xing F, Chen LL (2014b) CRISPR-P: a web tool for synthetic single guide RNA design of CRISPR-system in plants. Mol Plant 7:1494–1496

Li M, Li X, Zhou Z, Wu P, Fang M, Pan X, Lin Q, Luo W, Wu G, Li H (2016a) Reassessment of the four yield-related genes Gn1a, DEP1, GS3, and IPA1 in rice using a CRISPR/Cas9 system. Front Plant Sci 7:377

Li T, Liu B, Chen CY, Yang B (2016b) TALEN-mediated homologous recombination produces site-directed DNA base change and herbicide-resistantrice. J Genet Genomics 43:297–305
Li J, Meng X, Zong Y, Chen K, Zhang H, Liu J et al (2016c) Gene replacements and insertions in rice by intron targeting using CRISPR-Cas9. Nat Plants 12:16139
Li M, Li X, Zhou Z, Wu P, Fang M, Pan X et al (2016d) Reassessment of the four yield-related genes Gn1a, DEP1, GS3, and IPA1 in rice using a CRISPR/Cas9 system. Front Plant Sci 7:377
Li X, Zhou W, Ren Y, Tian X, Lv T, Wang Z, Fang J, Chu C, Yang J, Bu Q (2017a) High-efficiency breeding of early-maturing rice cultivars via CRISPR/Cas9-mediated genome editing. J Genet Genomics 44:175–178
Li X, Zhou W, Ren Y, Tian X, Lv T, Wang Z et al (2017b) High-efficiency breeding of early-maturing rice cultivars via CRISPR/Cas9-mediated genome editing. J Genet Geno 44:175–178
Li J, Sun Y, Du J, Zhao Y, Xia L (2017c) Generation of targeted point mutations in rice by a modified CRISPR/Cas9 system. Mol Plant 10:526–529
Li X, Wang Y, Chen S, Tian H, Fu D, Zhu B, Luo Y, Zhu H (2018) Lycopene is enriched in tomato fruit by CRISPR/Cas9-mediated multiplex genome editing. Front Plant Sci 9:559
Li R, Liu C, Zhao R, Wang L, Chen L, Wenqing Yu, Zhang S, Sheng J, Shen L (2019) CRISPR/Cas9-Mediated SlNPR1 mutagenesis reduces tomato plant drought tolerance. BMC Plant Biol 19:38
Liang Z, Zhang K, Chen K, Gao C (2014) Targeted mutagenesis in Zea mays using TALENs and the CRISPR/Cas system. J Genet Genomics 41:63–68
Liang G, Zhang H, Lou D, Yu D (2016) Selection of highly efficient sgRNAs for CRISPR/Cas9-based plant genome editing. Sci Rep 6:21451
Liu D, Chen X, Liu J, Ye J, Guo Z (2012) The rice ERF transcription factor OsERF922 negatively regulates resistance to *Magnaporthe oryzae* and salt tolerance. J Exp Bot 63:3899–3912
Liu H, Wei Z, Dominguez A, Li Y, Wang X, Qi LS (2015) CRISPR-ERA: a comprehensive design tool for CRISPR-mediated gene editing, repression and activation. Bioinform 31(22):3676–3678
Lombardo L, Coppola G, Zelasco S (2016) New technologies for insect-resistant and herbicide-tolerant plants. Trends Biotechnol 34:49–57
Lou D, Wang H, Liang G, Yu D (2017) OsSAPK2 confers abscisic acid sensitivity and tolerance to drought stress in rice. Front Plant Sci 8:993
Lowder LG, Zhang D, Baltes NJ, Paul JW, Tang X, Zheng X et al (2015) A CRISPR/Cas9 toolbox for multiplexed plant genome editing and transcriptional regulation. Plant Physiol 169:971–985
Ma X, Liu YG (2016) CRISPR/Cas9-based genome editing systems and analysis of targeted genome mutations in plants. Hereditas (Beijing) 38:118–125
Ma M, Ye AY, Zheng W, Kong L (2013) A guide RNA sequence design platform for the CRISPR/Cas9 system for model organism genomes. BioMed Res Int 270805
Ma M, Zhao H, Li Z, Hu S, Song W, Liu X, Leubner G (2016) Retracted: TaCYP78A5 regulates seed size in wheat (Triticum aestivum). J Exp Bot 67(5):1397–1410
Ma J, Chen J, Wang M, Ren Y, Wang S, Lei C, Cheng Z, Sodmergen (2018) Disruption of OsSEC3A increases the content of salicylic acid and induces plant defense responses in rice. J Exp Bot 69:1051–1064
Makhotenko AV, Khromov AV, Snigir EA, Makarova SS, Makarov VV, Suprunova TP, Kalinina NO, Taliansky ME (2019) Functional analysis of coilin in virus resistance and stress tolerance of potato solanum tuberosum using CRISPR-Cas9 Editing. Doklady Biochem Biophys 484(1):88–91
Malnoy M, Jin Q, Borejsza-Wysocka EE, He SY, Aldwinckle HS (2007) Overexpression of the Apple MpNPR1 gene confers increased disease resistance in Malus × domestica. Mol Plant Microbe In 20(12):1568–1580
Mao X, Zheng Y, Xiao K, Wei Y, Zhu Y, Cai Q et al (2018) OsPRX2 contributes to stomatal closure and improves potassium deficiency tolerance in rice. Biochem Biophys Res Commun 495:461–467
Miao C, Xiao L, Hua K, Zoua C, Zhao Y, Bressanb RA et al (2018) Mutations in a subfamily of abscisic acid receptor genes promote rice growth and productivity. Proc Natl Acad Sci USA 115:6058–6063

Montague TG, Cruz JM, Gagnon JA, Church GM, Valen E (2014) CHOPCHOP: a CRISPR/Cas9 and TALEN web tool for genome editing. Nucleic Acids Res

Moreno-Mateos MA, Vejnar CE, Beaudoin JD, Fernandez JP, Mis EK, Khokha MK, Giraldez AJ (2015) CRISPRscan: designing highly efficient sgRNAs for CRISPRCas9 targeting in vivo. Nat Methods (advance online publication)

Naito Y, Hino K, Bono H, Ui-Tei K (2014) CRISPRdirect: software for designing CRISPR/Cas guide RNA with reduced off-target sites. Bioinformatics

Narayanan A, Hill-Teran G, Moro A, Ristori E, Kasper DM, Roden CA et al (2016) In vivo mutagenesis of miRNA gene families using a scalable multiplexed CRISPR/Cas9 nuclease system. Sci Rep 6:32386

Nekrasov V, Wang C, Win J, Lanz C, Weigel D, Kamoun S (2017) Rapid generation of a transgene-free powdery mildew resistant tomato by genome deletion. Sci Rep 7:482

O'Brien A, Bailey TL (2014) GT-scan: identifying unique genomic targets. Bioinformatics 30:2673–2675

Odipio J, Alicai T, Ingelbrecht I, Nusinow DA, Bart R, Taylor NJ (2017) Efficient CRISPR/Cas9 genome editing of phytoene desaturase in cassava. Front Plant Sci 8:1780

Osakabe Y, Watanabe T, Sugano SS, Ueta R, Ishihara R, Shinozaki K et al (2016) Optimization of CRISPR/Cas9 genome editing to modify abiotic stress responses in plants. Sci Rep 6:26685

Ozseyhan ME, Kang J, Mu X, Lu C (2018) Mutagenesis of the FAE1 genes significantly changes fatty acid composition in seeds of *Camelina sativa*. Plant Physiol Bioch 123:1–7

Pandey P, Irulappan V, Bagavathiannan MV, Kumar MS (2017) Impact of combined abiotic and biotic stresses on plant growth and avenues for crop improvement by exploiting physio-morphological traits. Front Plant Sci 8:537

Peng A, Shanchun C, Tiangang L, Lanzhen X, Yongrui H, Liu W et al (2017) Engineering canker-resistant plants through CRISPR/Cas9-targeted editing of the susceptibility gene CSLOB1 promoter in citrus. Plant Biotechnol J 15:1509–1519. https://doi.org/10.1111/pbi.12733

Prykhozhij SV, Rajan V, Gaston D, Berman JN (2015) CRISPR MultiTargeter: a web tool to find common and unique CRISPR single guide RNA targets in a set of similar sequences. PLoS ONE 10:e0119372

Pyott DE, Sheehan E, Molnar A (2016a) Engineering of CRISPR/Cas9-mediated potyvirus resistance in transgene-free *Arabidopsis* plants. Mol Plant Pathol 17:1276–1288

Pyott DE, Sheehan E, Molnar A (2016b) Engineering of CRISPR/Cas9mediated potyvirus resistance in transgene-free Arabidopsis plants. Mol Plant Pathol 17:1276–1288

Ricroch A, Clairand P, Harwood W (2017) Use of CRISPR systems in plant genome editing: toward new opportunities in agriculture. Emerg Top Life Sci 1:169–182

Rodríguez-Leal D, Lemmon ZH, Man J, Bartlett ME, Lippman ZB (2017) Engineering quantitative trait variation for crop improvement by genome editing. Cell 171:470–480

Sánchez-León S, Gil-Humanes J, Ozuna CV, Giménez MJ, Sousa C, Voytas DF, Barro F (2018) Low-gluten, nontransgenic wheat engineered with CRISPR/Cas9. Plant Biotechnol J 16:902–910

Sauer NJ, Jerry M, Miller RB, Warburg ZJ, Walker KA, Beetham PR et al (2016) Oligonucleotide-directed mutagenesis for precision gene editing. Plant Biotechnol J 14:496–502

Savary S, Ficke A, Aubertot JN, Hollier C (2012) Crop losses due to diseases and their implications for global food production losses and food security. Food Sec 4:519–537

Shan Q, Wang Y, Li J, Zhang Y, Chen K, Liang Z et al (2013) Targeted genome modification of crop plants using a CRISPR-Cas system. Nat Biotechnol 31:686–688

Shan Q, Zhang Y, Chen K, Zhang K, Gao C (2015) Creation of fragrant rice by targeted knockout of the OsBADH2 gene using TALEN technology. Plant Biotechnol J 13:791–800

Shao G, Xie L, Jiao G, Wei X, Sheng Z, Tang S et al (2017) CRISPR/CAS9 mediated editing of the fragrant Gene Badh2 in Rice. Chin J Rice Sci 31:216–222

Shen R, Wang L, Liu X, Wu J, Jin W, Zhao X, Chen L (2017) Genomic structural variation-mediated allelic suppression causes hybrid male sterility in rice. Nature Commun 8(1):1–10

Shi J, Gao H, Wang H, Lafitte HR, Archibald RL, Yang M, Hakimi SM, Mo H, Habben JE (2017) ARGOS8 variants generated by CRISPR-Cas9 improve maize grain yield under field drought stress conditions. Plant Biotechnol J 15(2):207–216

Shimatani Z, Kashojiya S, Takayama M, Terada R, Arazoe T, Ishii H et al (2017) Targeted base editing in rice and tomato using a CRISPR-Cas9 cytidine deaminase fusion. Nat Biotechnol 35:441–443

Shukla VK, Doyon Y, Miller JC, Dekelver RC, Moehle EA, Worden SE et al (2009) Precise genome modification in the crop species *Zea mays* using zinc-finger nucleases. Nature 459:437

Song G, Jia M, Chen K, Kong X, Khattak B, Xie C et al (2016) CRISPR/Cas9: a powerful tool for crop genome editing. Crop J 4:75–82

Soyk S, Muller NA, Park SJ, Schmalenbach I, Jiang K, Hayama R, Zhang L et al (2017) Variation in the flowering gene Self Pruning 5G promotes day-neutrality and early yield in tomato. Nat Genet 49:162–168

Stemmer M, Thumberger T, del Sol KM, Wittbrodt J, Mateo JL (2015) CCTop: an intuitive, flexible and reliable CRISPR/Cas9 target prediction tool. PLoS ONE 10:e0124633

Sun Y, Zhang X, Wu C, He Y, Ma Y, Hou H et al (2016a) Engineering herbicide-resistant rice plants through CRISPR/Cas9-mediated homologous recombination of Acetolactate synthase. Mol Plant 9:628–631

Sun Y, Zhang X, Wu C, He Y, Ma Y, Hou H et al (2016b) Engineering herbicide-resistant rice plants through CRISPR/Cas9-mediated homologous recombination of acetolactate synthase. Mol Plant 9:628–631

Sun Y, Jiao G, Liu Z, Zhang X, Li J, Guo X et al (2017) Generation of high-amylose rice through CRISPR/Cas9-mediated targeted mutagenesis of starch branching enzymes. Front Plant Sci 8:298

Svitashev S, Young J, Schwartz C, Gao H, Falco SC, Cigan AM (2015) Targeted mutagenesis, precise gene editing and site-specific gene insertion in maize using Cas9 and guide RNA. Plant Physiol 169:931–945

Tang X, Lowder LG, Zhang T, Malzahn AA, Zheng X, Voytas DF et al (2017) A CRISPR-Cpf1 system for efficient genome editing and transcriptional repression in plants. Nat Plants 3:17018

Taylor JE, Hatcher PE, Paul ND (2004) Crosstalk between plant responses to pathogens and herbivores. J Exp Bot 55:159–168

Townsend JA, Wright DA, Winfrey RJ, Fu F, Maeder ML, Joung JK et al (2009) High-frequency modification of plant genes using engineered zinc-finger nucleases. Nature 459:442

Ueta R, Abe C, Watanabe T, Sugano SS, Ishihara R, Ezura H et al (2017) Rapid breeding of parthenocarpic tomato plants using CRISPR/Cas9. Sci Rep 7:507

Vats S, Kumawat S, Kumar V, Patil GB, Joshi T, Sonah H, Sharma TR, Deshmukh R (2019) Genome editing in plants: exploration of technological advancements and challenges. Cells 8(11):1386

Wally O, Jayaraj Zamir J, Punja ZK (2009) Broad-spectrum disease resistance to necrotrophic and biotrophic pathogens in transgenic carrots (*Daucus carota* L.) expressing an Arabidopsis NPR1 gene. Planta 231(1):131–141

Waltz E (2016) CRISPR-edited crops free to enter market, skip regulation. Nat Biotechnol 34:582

Wang Y, Cheng X, Shan Q, Zhang Y, Liu J, Gao C, Qiu JL (2014a) Simultaneous editing of three homoeoalleles in hexaploid bread wheat confers heritable resistance to powdery mildew. Nat Biotechnol 32:947–951

Wang Y, Cheng X, Shan Q, Zhang Y, Liu J, Gao C et al (2014b) Simultaneous editing of three homoeoalleles in hexaploid bread wheat confers heritable resistance to powdery mildew. Nat Biotechnol 32:947–951

Wang F, Wang C, Liu P, Lei C, Hao W, Gao Y et al (2016a) Enhanced rice blast resistance by CRISPR/Cas9-targeted mutagenesis of the ERF transcription factor gene OsERF922. PLoS ONE 11:0154027

Wang F, Wang C, Liu P, Lei C, Hao W, Gao Y et al (2016b) Enhanced rice blast resistance by CRISPR/Cas9-targeted mutagenesis of the Ezrf transcription factor gene OSERF922. PLoS ONE 11:e0154027

Wang L, Chen L, Li R, Zhao R, Yang M, Sheng J, Shen L (2017a) Reduced drought tolerance by CRISPR/Cas9-mediated SlMAPK3 mutagenesis in tomato plants. J Agr Food Chem 65:8674–8682

Wang L, Chen L, Li R, Zhao R, Yang M, Sheng J et al (2017b) Reduced drought tolerance by CRISPR/Cas9-mediated SlMAPK3 mutagenesis in tomato plants. J Agric Food Chem 65:8674–8682

Wang X, Tu M, Wang D, Liu J, Li Y, Li Z, Wang Y, Wang X (2018) CRISPR/Cas9-mediated efficient targeted mutagenesis in grape in the first generation. Plant Biotechnol J 16:844–855

Watanabe K, Oda-Yamamizo C, Sage-Ono K, Ohmiya A, Ono M (2018) Alteration of flower colour in Ipomoea nil through CRISPR/Cas9-mediated mutagenesis of carotenoid cleavage dioxygenase 4. Transgenic Res 27(1):25–38

Weeks DP, Spalding MH, Yang B (2015) Use of designer nucleases for targeted gene and genome editing in plants. Plant Biotechnol J 14:483–495

Wong N, Liu W, Wang X (2015) WU-CRISPR: characteristics of functional guide RNAs for the CRISPR/Cas9 system. Genome Biol 16:218

Xiao A, Cheng Z, Kong L, Zhu Z, Lin S, Gao G, Zhang B (2014) CasOT: a genome-wide Cas9/gRNA off-target searching tool. Bioinformatics

Xie K, Yang Y (2013) RNA-guided genome editing in plants using a CRISPR-Cas system. Mol Plant 6:1975–1983

Yao L, Zhang Y, Liu C, Liu Y, Wang Y, Liang D, Liu J, Sahoo G, Kelliher T (2018) OsMATL mutation induces haploid seed formation in indica rice. Nat Plants 4:530–533

Xie S, Shen B, Zhang C, Huang X, Zhang Y (2014) sgRNAcas9: a software package for designing CRISPR sgRNA and evaluating potential off-target cleavage sites. PLoS ONE 9:e100448

Xing Y, Zhang Q (2010) Genetic and molecular bases of rice yield. Annu Rev Plant Biol 61:421–442

Xu Z, Zheng Y, Zhu Y, Kong X, Hu L, Umen JG (2011) Evidence for OTUD-6B participation in B lymphocytes cell cycle after cytokine stimulation. PLoS ONE 6(1):e14514

Xu H, Xiao T, Chen CH, Li W, Meyer C, Wu Q, Wu D, Cong L, Zhang F, Liu JS, Brown M, Liu SX (2015) Sequence determinants of improved CRISPR sgRNA design. Genome Res

Xu R, Yang Y, Qin R, Li H, Qiu C, Li L, Wei P, Yang J (2016) Rapid improvement of grain weight via highly efficient CRISPR/Cas9-mediated multiplex genome editing in rice. J Genet Genomics 43:529–532

Yang H, Wu J, Tang T, Liu K, Dai C (2017) CRISPR/Cas9-mediated genome editing efficiently creates specific mutations at multiple loci using one sgRNA in *Brassica napus*. Sci Rep 7:7489

Yang Y, Zhu K, Li H, Han S, Meng Q, Khan SU, Fan C, Xie K, Zhou Y (2018) Precise editing of CLAVATA genes in *Brassica napus* L. regulates multilocular silique development. Plant Biotechnol J 16:1322–1335

Zhang H, Zhang J, Wei P, Zhang B, Gou F, Feng Z et al (2014) The CRISPR/Cas9 system produces specific and homozygous targeted gene editing in rice in one generation. Plant Biotechnol J 12:797–807

Zhang Y, Liang Z, Zong Y, Wang Y, Liu J, Chen K, Qiu J, Gao C (2016) Efficient and transgene-free genome editing in wheat through transient expression of CRISPR/Cas9 DNA or RNA. Nat Commun 7:12617

Zhao N, Xu X, Wamboldt Y, Mackenzie SA, Yang X, Hu Z et al (2016) MutS HOMOLOG1 silencing mediates ORF220 substoichiometric shifting and causes male sterility in Brassica juncea. J Exp Bot 67:435–444

Zhang Y, Bai Y, Wu G, Zou S, Chen Y, Gao C, Tang D (2017) Simultaneous modification of three homoeologs of TaEDR1 by genome editing enhances powdery mildew resistance in wheat. Plant J 91:714–724

Zhang Y, Li S, Xue S, Yang S, Huang J, Wang L (2018) Phylogenetic and CRISPR/Cas9 studies in deciphering the evolutionary trajectory and phenotypic impacts of rice ERECTA genes. Front Plant Sci 9:473

Zhao Y, Zhang C, Liu W, Gao W, Liu C, Song G et al (2016) An alternative strategy for targeted gene replacement in plants using a dual-sgRNA/Cas9 design. Sci Rep 6:23890

Zhou J, Peng Z, Long J, Sosso D, Liu B, Eom JS et al (2015) Gene targeting by the TAL effector PthXo2 reveals cryptic resistance gene for bacterial blight of rice. Plant J 82:632–643

Zhu LJ, Holmes BR, Aronin N, Brodsky MH (2014) CRISPRseek: a bioconductor package to identify target-specific guide RNAs for CRISPR-Cas9 genome-editing systems. PLoS ONE 9:e108424

Zou T, Liu M, Xiao Q, Wang T, Chen D, Luo T et al (2018) OsPKS2 is required for rice male fertility by participating in pollen wall formation. Plant Cell Rep 37:759–773

Chapter 9
Regulatory Framework and Policy Decisions for Genome-Edited Crops

Anirudh Kumar, Rakesh Kumar, Nitesh Singh, and Aadil Mansoori

Abstract Gene editing methods have been considered as one of the important approaches that have a huge impact on plant breeding. Currently, gene editing is considered as one of the most powerful techniques for crop improvement, which offers precise base editing in a predictable manner. Within a short span of time, genome editing based breeding has made remarkable advances in the area of crop improvement as well as on human lives, due to cost-effectiveness, reliability and easy to use. However, it has also raised plethora of ethical and legal debates among research organizations, funding agencies besides national and international policy makers. The bigger question is do the prevailing law, directives and regulations anticipate and/or address the societal concern about genome-edited crop? Here, we review the existing regulatory framework and ethical complications that may influence the adoption of technology and its acceptance. Additionally, perspective for an alternative regulatory system that appropriately matches the scientific progress is discussed.

Keywords Gene editing · Genome editing · Directives · Regulation · CRISPR/Cas · Crop improvement · Plant breeding

9.1 Introduction

Re-orientation of crop improvement strategy is a prerequisite to meet the requirement of an ever-growing global population under limiting resources and predicted climate change implications. Genetic gains achieved through plant breeding (crossing and

A. Kumar (✉) · N. Singh · A. Mansoori
Department of Botany, Indira Gandhi National Tribal University (IGNTU), Amarkantak 484887, Madhya Pradesh, India
e-mail: anirudh.kumar@igntu.ac.in

R. Kumar
Department of Life Science, Central University of Karnataka, Kalaburagi, Karnataka, India

A. Bhattacharya et al. (eds.), *CRISPR/Cas Genome Editing*, Concepts and Strategies in Plant Sciences, https://doi.org/10.1007/978-3-030-42022-2_9

selection) has made a substantial contribution in terms of development and dissemination of high yielding resilient crops (Ejeta 2009). However, the crossing and selection face limitations with polyploidy, heterozygosity, self-incompatibility and longer generation time. Subsequently, genetic variations through mutation breeding were adopted where selection and screening remain time-consuming and expensive. Mutation breeding is regarded as non-transgenic or non-genetic engineering technique and hence, was exempted from GM legislation annexure of Directive 2001/18/EC (Gracia 2005). However, mutation breeding through chemical or physical mutagens was not as successful as expected; mainly due to (i) low frequency of non-synonymous mutations, (ii) require extensive backcrossing for the removal of non-essential mutations (background mutations acquired due to mutagen), and (iii) crop ploidy remained a challenge as higher ploidy level reduces the chance of effective phenotype as well as required more number of individuals in a population to identify homozygous line.

Scientific progress and novelty have been always implemented to facilitate plant breeding in order to improve precision in genome alteration. Genome or gene editing is an addition in the continuum of plant breeding innovation. It allows to create precise genetic variation (exchange, insertion, deletion) within an existing gene pool. Genome editing techniques include protein-mediated techniques (viz. TALENS, ZFN), nucleic acid-mediated techniques (viz. ODM) and combination of these two (viz. CRISPR). It offers great opportunities for crop improvement and could lead to gene stacking (Puchta 2017) but it also created regulatory challenges across the world. In the polarised world, some countries have developed the mechanism to address the issue of regulatory status of genome-edited crops. However, the legal status remains unclear across many countries particularly, in case of CRISPR/Cas derived crops (Whelan and Lema 2015; Wolt et al. 2016). In the year 2007, European commission instituted "New Techniques Working Group (NTWG)" to evaluate new techniques of GM with reference to prevailing European Union (EU) legislation on genetically modified organism (GMO) (Lusser et al. 2012). The legal opinion made by NTWG has never been officially released. However, European Commission has exempted mutagenesis-derived plants from the directive, if no recombinant nucleic acid is incorporated.

In the USA, specific GMO legislation was not constituted. However, existing legislation which was taking care of plant protection, pest management and food safety was used to regulate the GMO (USDA 2017). Canada has set up special regulatory provision for the "plant with novel traits (PNTs)" where any technology derived plants have not been kept under legislation, rather development of novel traits trigger legislation (Custers 2017). However, the prevailing EU GM legislation (European Directive 2001/18/EC) is process-based provoked by the implementation of GM steps. As far as Germany is concerned, the BVL (Federal Office of Consumer Protection and Food Safety) do not consider CRISPR derived crop as recombinant. Similarly, the Swedish Board of Agriculture (SBA) considers CRISPR/Cas like mutagenesis provided no foreign DNA occurs in the final plant/product. Nevertheless, due to lack of data about the long-term effect and uniform guidelines across the world on the legal status of genome-edited plants. Till now, lack of clarity exists on the regulatory framework of genome-edited crops. There is no clear-cut demarcation that the regulatory framework should be executed on the process adopted (process-based approach) or product generated (product-based approach). Concurrently, whether the same legal status of crops generated through point mutation and spontaneous mutation will be followed or different, is also not crystal clear and hence, is a matter of discussion. Moreover, asynchronous regulations must be synchronized for satisfactory risk assessment. The current chapter has made an effort to highlight the provisions of directives, ethical issues as well as regulatory framework related to genome-edited crops.

9.2 Regulatory Scenario

Regulatory framework was developed to monitor the GM crops more than two decades ago on the basis of limpid distinction between transgenics products and conventional breeding products. However, dearth of continuum between genetic engineering and conventional plant breeding was observed at the time of prevailing EU directive (Directive 2001/18/EC).

According to EU legislation, the definition of GM crops is "any organism whose genetic material has been altered in such a way that does not originate by natural mating and/or natural recombination". However, techniques involving nucleic acid were kept under the surveillance of regulation whereas mutagenesis was kept out of the regulations (European Parliament 2001). Thus, crops derived from genetic alterations were subject to follow stringent regulatory guidelines, while mutagenesis derived crops were excluded from EU regulatory guidelines. This stringent guideline somehow retarded the entry of GM crops into the market (Smyth 2014). Slow approval and long queue of pending cases within the EU countries further restricted the rate of commercialization of GM crops (European Commission 2015). It seems that the EU precautionary measures and genetic modification defined process increases the complexity of regulatory framework across the world (Bayer et al. 2010; Okeno et al. 2013). Other countries like Australia, Argentina and Brazil have positively instigated process-based regulatory approach which is relatively consistent compared to North America (Smyth and Phillips 2014). To date, even a single internationally agreed regulatory framework do not exist; nevertheless, there are many countries that are in the process of evaluation whether and/or to what extent the existing regulations are necessary for research work and to access the products of gene editing.

9.3 Current Status and Opinion

Various scientific communities and regulatory policy makers have reviewed the status of existing regulations of genome-edited crops (Breyer et al. 2009; Lusser and Davies 2013; Pauwels et al. 2014). There is no consensus, however, nature of DNA repair process, usage of phenotype developed and existing regulatory framework to release product are the focus point of discussion across the globe. EU scientists are exploring genome editing with engineered nucleases (GEEN) and new plant breeding technology for crop improvement and to avoid strict GM regulatory framework (Hartung and Schiemann 2014). There is hope within EU community of paradigm shift of product over process (Hartung and Schiemann 2014). Currently, the main objective is to ensure that GEEN and other technology which uses site-directed mutation should not fall under any regulation across the world, exception is Canada and its PNT regulation (International Life Science Institute 2013). With the emergence of genome editing techniques applied in crop development and regulatory framework, the three broad categories have been framed. The category 1 technique involves transient integration of rDNA into host genome (which does not integrate into plant genome) with subsequent removal (Pauwels et al. 2014). There are three famous examples of Category 1 technique, which includes site-specific point mutation, site-specific random mutation and site-specific mutation with oligonucleotides (OMM), by NHEJ (SDN1) and with DNA repair (SDN2) respectively (Wolt et al. 2016). The category 2 technique involves stable integration of rDNA into host genome. It also includes the intermediate steps involving expression of SDN1 and SDN2 (Lusser and Davies 2013). The category 3 technique involves stable integration of rDNA into host genome. In this category, GEEN technology is used to deliver the transgene/s through infusion of homologous recombination (SDN3). Site-directed stacking of transgene fall under this category (D'Halluin et al. 2013).

On the basis of various methodologies used for genome editing, it can be demonstrated that the regulatory consideration is a never-ending process across the world. Usually, decisions are made on the merits of the case through case by case study. Therefore, until the development of consensus among regulatory bodies and single uniform regulatory guideline, there will be uncertainty. Today there is scientific sentiment for different regulatory standards for gene-edited crops and transgenic crops. However, some NGO groups argue otherwise and show over-concern and do not want to give any regulatory benefits to gene-edited crops (Camacho et al. 2014). Finally, it can be concluded that plant developed through GEEN techniques will face various regulatory constraints based on interpretation of an obscure regulatory framework. In USA, USDA has specified a guideline to product developers which recommend that site-directed approaches leading to targeted deletion of endogenous nucleotide (SDN1) and transgenesis intermediary steps used to check the absence of transgenic element would not be regulated articles under USDA status, however, in case of site-directed oligonucleotide insertions or substitution, case-specific approval will be required (Camacho et al. 2014). In fact, the product derived from these new plant breeding technologies (NPBT) including genome-edited/GMOs are now in market because of strong support USDA (Fig. 9.1).

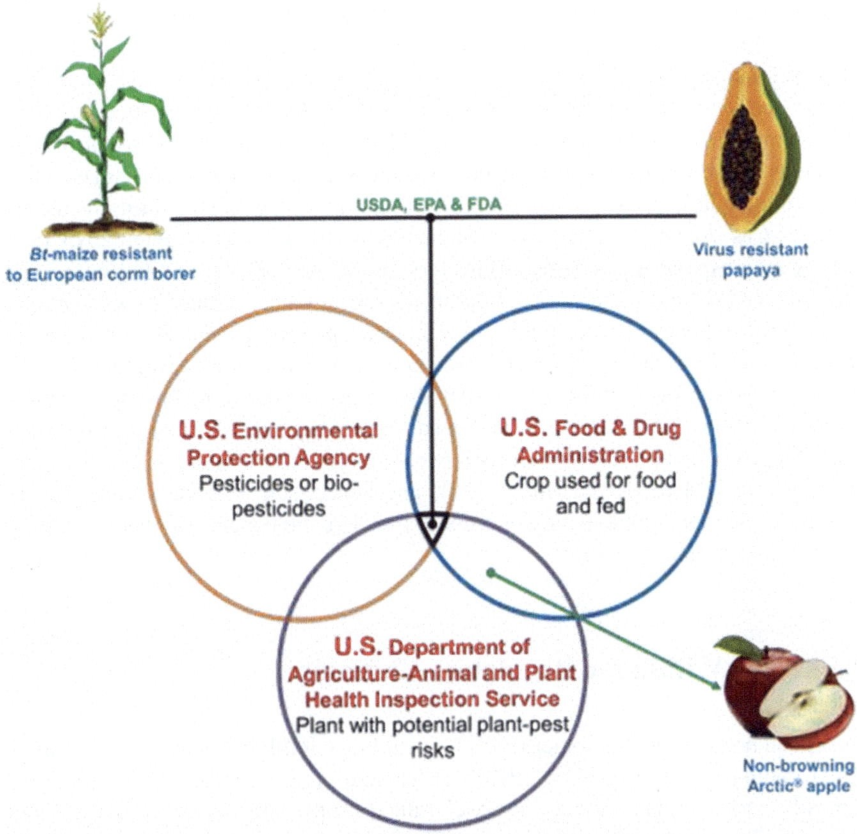

Fig. 9.1 The role of regulatory agencies in US for the approval of GMOs/Genome-edited crop and some examples of approved products. *Source* Original source Turner 2014; modified and adopted from a book chapter (https://www.ncbi.nlm.nih.gov/books/NBK424533/ or Doi: https://doi.org/10.17226/23395)

9.4 Public Understanding and Regulatory Implications

Plant biologists are constantly working towards improvising genome editing method which may get better social and regulatory acceptance over transgenic approaches. Today the biggest challenge is not only to develop the best scientific and technical methods; rather it is gaining confidence of public and regulatory authorities for easy approval of the edited crops (Chapoti and Wolt 2007). The restrictive regulations impact both product developers and public desire. The ability of US regulatory authority with regard to new breeding technologies appears challengeable, while their track record is exemplary well when it comes to transgenic regulation and safety (Camacho et al. 2014). Greater openness by US in the regulation of genome editing and innovation has outpaced Europe, who has led in the area till 2011. The major reason behind this setback was the incapability of EU to make an advancement in the adoption of GM crops and delayed in taking a position on regulation of genome editing (Funk 2015).

The general public opinion is not encouraging as far as food produced through biotechnological intervention is concerned. This has led to creating a hurdle that has delayed the implementation of the regulatory process. There is a huge difference among public opinions and scientific community on genome editing including CRISPR/Cas. According to a survey conducted by Pew Research Center, 2015, only 37% people responded positively on GM food; however 88% of scientific community recognizes GM food as safe. The greater discrepancy among public and scientific community indicate that the public may not accept food produced from genome editing technology unless transparent science-based campaign is conducted early. In fact, a section of scientist also argues that genome editing technology requires rigorous scrutiny in contrast to established technology such as transgenics (Araki and Ishii 2015). Additionally, civil society campaigns have also created doubt about the safety of GM foods (Paarlberg 2014) and hence, scientific community has to communicate more emphatically in order to translate the benefits of genome editing technology for greater purposes.

9.5 Need Within the Regulatory Community

New plant breeding technologies (NPBT) including CRISPR provides feasible alternative compared to transgenics. CRISPR offers new opportunities for innovation and it may be very useful in introducing novel traits in crops. The success of CRISPR and other GEEN technology will be limited like GMO if public opinion is not changed and regulatory procesess are broken in many parts of the world. To harness the benefit from these modern breeding technologies, the regulatory processes have to be made easy and uniform globally. The continued belief in process-based definitions and process-based language in public discourse reduced the aptly assessed approaches for the genome-edited crops. This will lead to public misunderstanding for crops derived from genome editing, which may create hurdle for regulators. Further, regulators may face pressure to evaluate the genome-edited crops within the existing biosafety framework. Thus, the focus for generation of novel phenotype/trait or trait staking is lost because of strict regulatory rules especially in the EU and other developing countries, resulted in closer of several research R&D units of the private seed companies which were focussed for improvisation of elite cultivars through NPBT. Fortunately, progress is being made across the globe by regulators in creating sensible, viable and pragmatic approaches towards implementation of genome editing tools for crop improvement. However, down the line product-based regulation should come for NPBT for a better world.

9.6 Conclusions and Perspectives

Although, the introduction of genome editing in plant breeding program has various benefits, however, the public observation and perception play a major role in commercialization of its product (Scheben et al. 2018). GMO food has lacked widespread public acceptance because of negative perception especially due to the public unawareness of the modern NPBT. Therefore, precaution has to be taken care of to avoid such negative public image about genome-edited crops (Frewer et al. 2004). Negative perception may create pressure on government to restrict the use of genome-edited crops, which may further limit the scientific innovation (Malyska et al. 2016). Therefore, public should be engaged in a honest dialogue regarding safety of genome-edited crops/foods. Public primary concern is the introduction of transgene which is not found in genome-edited crops; thus the chance of public acceptance is very high. There is also need for transparent legislation which can cover existing and future genome-edited based plant breeding program. The inconsistency in regulation of chemical and radiation mutagenesis which hold greater risk compared to genome editing (Urnov et al. 2018). Importantly, the potential risk of genome editing must be evaluated in conjunction with the benefits that the technology is expected to carry.

Acknowledgements RK acknowledges Science and Engineering Research Board (SERB), Department of Science and Technology (DST), Government of India for the financial support. A.K sincerely thanks University Grants Commission (UGC) start-up grant (No.F.30-392/2017 (BSR) and Madhya Pradesh Council of Science and Technology (Endt. No. 3879/CST/R&D/BioSci/2018) for the funding to the laboratory.

Ethics Approval and Consent to Participate
Not applicable.

Competing Interests

The authors declare that they have no competing interests.

Author Contributions

AK, RK, NS and AM wrote the MS. All authors have read and approved the MS for publication.

References

Araki M, Ishii T (2015) Towards social acceptance of plant breeding by genome editing. Trends Pl Sci 20:145–149

Bayer JC, Norton GW, Falck-Zepeda JB (2010) Cost of compliance with biotechnology regulation in the Philippines: implications for developing countries

Breyer D, Herman P, Brandenburger A, Gheysen G, Remaut E et al (2009) Commentary: genetic modification through oligonucleotide-mediated mutagenesis. A GMO regulatory challenge? Environmental Biosafety Res 8:57–64

Camacho A, Van Deynze A, Chi-Ham C, Bennett AB (2014) Genetically engineered crops that fly under the US regulatory radar. Nature Biotech 32:1087

Chapoti SM, Wolt JD (2007) Genetically modified crops for the bioeconomy: meeting public and regulatory expectations. Transgenic Res 16:675–688

Custers R (2017) The regulatory status of gene-edited agricultural products in the EU and beyond. Emerg Topics Life Sci 1:221–229

D'Halluin K, Vanderstraeten C, Van Hulle J, Rosolowska J, Van Den Brande I, Pennewaert A, Broadhvest J (2013) Targeted molecular trait stacking in cotton through targeted double-strand break induction. Plant Biotech J 11:933–941

Ejeta G (2009) Revitalizing agricultural research for global food security. Food Security 1:391

European Commission (2015) Proposal for a regulation of the European Parliament and of the Council amending regulation (EC) No 1829/2003 as regards the possibility for the Member States to restrict or prohibit the use of genetically modified food and feed on their territory

European Parliament (2001) Directive 2001/18/EC of the European Parliament and of the Council of 12 March 2001 on the deliberate release into the environment of genetically modified organisms and repealing Council Directive 90/220/EEC–Commission Declaration. Official J Euro Union 40:1–39

Frewer L, Lassen J, Kettlitz B, Scholderer J, Beekman V, Berdal KG (2004) Societal aspects of genetically modified foods. Food Chem Toxicol 42:1181–1193

Funk C, Rainie L (2015) Public and scientists' views on science and society. Pew Research Center 29

García PR (2005) Directive 2001/18/EC on the Deliberate release into the environment of GMOs: an overview and the main provisions for placing on the market. Policy 263

Hartung F, Schiemann J (2014) Precise plant breeding using new genome editing techniques: opportunities, safety and regulation in the EU. Plant J 78:742–752

International Life Sciences Institute (2013) ILSI workshop on new breeding technologies (NBT). International Life Sciences Institute SEA Region Australasia, Canberra

Lusser M, Davies HV (2013) Comparative regulatory approaches for groups of new plant breeding techniques. New Biotech 30:437–446

Lusser M, Parisi C, Plan D, Rodríguez-Cerezo E (2012) Deployment of new biotechnologies in plant breeding. Nat Biotech 30:231

Malyska A, Bolla R, Twardowski T (2016) The role of public opinion in shaping trajectories of agricultural biotechnology. Trends Biotech 34:530–534

National Academies of Sciences, Engineering, and Medicine (2016) Genetically engineered crops: experiences and prospects. The National Academies Press, Washington, DC. https://doi.org/10.17226/23395

Okeno JA, Wolt JD, Misra MK, Rodriguez L (2013) Africa's inevitable walk to genetically modified (GM) crops: opportunities and challenges for commercialization. New Biotech 30:124–130

Paarlberg R (2014) A dubious success: the NGO campaign against GMOs. GM Crops Food 5:223–228

Pauwels K, Podevin N, Breyer D, Carroll D, Herman P (2014) Engineering nucleases for gene targeting: safety and regulatory considerations. New Biotechnol 31:18–27

Puchta H (2017) Applying CRISPR/Cas for genome engineering in plants: the best is yet to come. Curr Opin Plant Biol 36:1–8

Scheben A, Edwards D (2018) Towards a more predictable plant breeding pipeline with CRISPR/Cas-induced allelic series to optimize quantitative and qualitative traits. Curr Opin Pl Biol 45:218–225

Smyth SJ (2014) The state of genetically modified crop regulation in Canada. GM Crops Food 5:195–203

Smyth SJ, Phillips PW (2014) Risk, regulation and biotechnology: the case of GM crops. GM Crops Food 5:170–177

Turner J (2014) Regulation of genetically engineered organisms at USDA–APHIS. Presentation to the National Academy of Sciences' Committee on Genetically Engineered Crops: Past Experience and Future Prospects, December 10. Washington, DC

Urnov FD, Ronald PC, Carroll D (2018) A call for science-based review of the European court's decision on gene-edited crops. Nat Biotech 36:800

Whelan AI, Lema MA (2015) Regulatory framework for gene editing and other new breeding techniques (NBTs) in Argentina. GM Crops Food 6:253–265

Wolt JD, Wang K, Yang B (2016) The regulatory status of genome-edited crops. Plant Biotech J 14:510–518

Chapter 10
Field Crop Improvement Using CRISPR/Cas9

Elangovan Mani

Abstract Though Agriculture contributes 6.4% of global GDP, it is vital for economic growth of any country. Plant breeders were the drivers to increase food production, through advanced hybrids and varieties, to meet the growing need of global population. Breeders relied on existing wild germplasm and created new germplasm through several mutation strategies to improve yield among several other traits like disease resistance. Mutation breeding is considered as the best source for creating variation through traditional breeding and has resulted in 3,200 improved crop varieties in more than 175 plant species. But, recently CRISPR based Gene editing is touted as the simple, technically low-cost, fast, and effective tool to induce point mutations, insertions/deletions (INDELs) of desired genes in specific genetic location. It has wider application in several key commercial polyploid field crops such as wheat, cotton, canola, sugarcane, tobacco, peanut, alfalfa, mustard, etc., and key diploid field crops like corn, soybean, rice, sunflower, sorghum, etc.

Keywords Field crops · CRISPR · Polyploids

10.1 Introduction

Green revolution and technology adaptations have played a significant role in the substantial increase in food production Worldwide; food security along with medical advancement has led to population growth in the developing countries in the second half of twentieth century (https://www.worldometers.info/world-population/). Green revolution has boosted the production of major crops such as rice, wheat, maize, sorghum, cotton, soybean, sunflower by transforming them into photo-insensitive, high yielding, and non-lodging varieties and hybrids. This has happened due to adopting important genetic mechanisms of development of photo-insentitivity, concentrated flowering (Florigen-antiflorgen action), and short plant types (Gibberillin A—DELLA) (Eshed and Lippman 2019). Wheat dwarfing genes Rht-B1b and

E. Mani (✉)
Advanta Seeds, UPL Ltd., Hyderabad, India
e-mail: elangene@gmail.com

A. Bhattacharya et al. (eds.), *CRISPR/Cas Genome Editing*, Concepts and Strategies in Plant Sciences, https://doi.org/10.1007/978-3-030-42022-2_10

Rht-D1b (which encodes mutant forms of DELLA proteins) and; Rice dwarf gene *sd-1* have produced the dwarfing phenotype due to deficiency in gibberellin (GA) plant growth hormones by a defective 20-oxidase GA biosynthetic enzyme (Spielmeyer et al. 2002). Both gene mutants were spontaneous, and thus promote the breeder's interest to develop new traits through artificial mutation introduction in the gene pool. Steady global population growth keeps plant breeders busy in innovating new ideas and tools to improve the yield potential of rice and other cereal crops. Targeting Induced Local Lesions IN Genomes (TILLING) is one of such tools that is being used for the production of mutant plant populations in crops, such as Sunflower, Rice, Barley, Soybean, Tomato and Wheat. In the last few years, the introduction of the concept of genome editing (GE) in crop plants has changed the mindset of agriculture scientists about gene pool enhancement and new trait development. GE technologies such as TALENs (Transcription activator-like effector nucleases), ZFN (Zinc Finger Nucleases) and CRISPR (Clustered regularly interspaced short palindromic repeats) accelerates the functional analyses of genes and the introduction of novel traits into important crop plants. The site-specific endonuclease-based systems, CRISPR genome editing technology has the potential to significantly speed up the development and commercialization of new traits in commercial row crops and specialty crops alike. Farming communities around the globe are facing several challenges such as unseasonal rains, fluctuations in commodity price due to globalization, high input cost, climate change, biotic and abiotic stresses, small farm holdings, economic disparity and political and regulatory differences between countries. This results into a consistent decline in the profit margin in agriculture. They are looking for affordable and climate-resilient crops to feed the ever-growing global population. The latest genome editing technologies have the potential to provide a solution for profitable and sustainable agriculture. This chapter briefly introduces current progress in the application of CRISPR-based gene editing tools in Field crops to develop commercially important traits of interest (Fig. 10.1).

10.2 Nutrition Improvement

In today's World, Agriculture productivity is emphasized on not only quantity, but the nutrition quality of food production. A recent report on 55 countries suggest that globally, the small farmers having landholding of <2 ha and constitutes 24% of gross agricultural land. They produce 28–31% of total crop production and manage 30–34% of the food supply (Ricciardi et al. 2018). Globally, there is an urgent need to improve the nutrition quality of the food, as current status of malnutrition in developing countries is alarming. GE technologies have the potential to make significant progress in the nutrition quality of cereal crops and other field crops and the CRISPR is the leading one.

Wheat: It is always a great help to mankind if healthy nutrition is brought out by the removal of compounds which causes an allergic reaction in Human. One

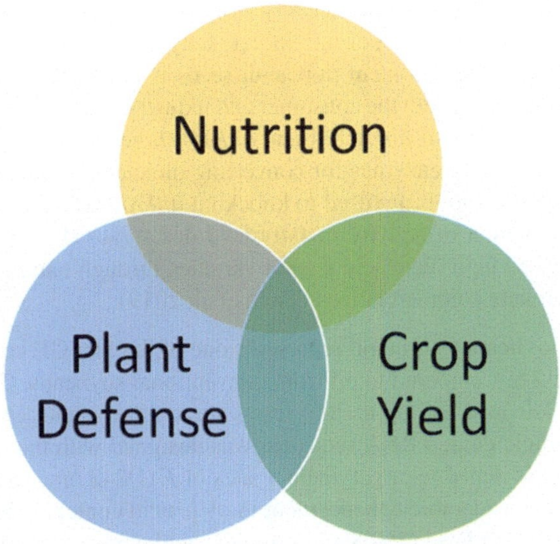

Fig. 10.1 The application of CRISPR based genome editing in various crops are catalogued based on three key purpose: Nutrition improvement, Plant defense-based traits, Yield improvement traits

of the widely reported allergic disorders, celiac disease is caused by proteins like gluten and 1–2% of humans are allergic to gluten. A gluten-free diet, excluding wheat, barley and rye, can provide relief to celiac patients. Short amino acid protein called gliadins and the long amino acid chain protein Glutenin makes the protein complex gluten in bread dough. The epitopes in α-, γ-, and ω- gliadins of glutenins trigger immunogenic reaction in celiac patients. Its presence is less in low molecular weight (LMW) glutenin while it is negligible in high molecular weight (HMW) which mostly safe. *Gli-2* locus (Chromosome 6) codes for α-gliadin genes, *Gli-1* locus (Chromosome 1) codes for 40 γ-gliadins, and *Gli-3* locus codes for 16 ω-gliadins. Though RNAi targeting all three gliadin families has successfully decreased 97% of gliadin expression in bread wheat grain, however, it is considered as transgenic alterations and needs to undergo genetically modified crop regulations. In a recent study, CRISPR is used as a tool to silence several genes coding family of gluten proteins. Six sgRNAs were designed, three of which target α-gliadins and the other three targeting γ-gliadins (Jouanin et al. 2019). Similar advancement can be expected for other nutritional advancements in wheat.

Peanut: Peanut is a good source of oil and proteins, about 1–2% of the population is affected by peanut allergies. Peanut has 13 genes responsible for allergens (*Ara h 1—Ara h 13*), which make up a significant proportion of the peanut's total protein content and some of them are critical for normal life cycle of the plants. For example, the concentration of *Ara h 1* in peanuts increases with kernel size, also, the protein expression of this gene is linked with peanut maturity (Brazelton 2015, Thesis). So engineering to develop hypoallergenic peanuts is challenging, but some biotech companies like Aranex Biotech, are closer to develop Hypoallergenic peanuts in future.

Developing countries rely more on oilseeds like a peanut to develop cooking oil. In routinely used peanut, linoleic fatty acid forms 70% of the total oil. Enhancement of oleic acid content in peanut seeds is one of the major goals in peanut breeding. This can benefit the consumer and industry, such as anti-oxidation and long shelf-life. Homeologous *ahFAD2A* and *ahFAD2B* genes express fatty acid desaturases, which are the key enzymes for converting oleic acid to linoleic acid that oxidizes readily. CRISPR tools are used to knockout *ahFAD2A* and *ahFAD2B* that led to a complete reduction of activity of fatty acid desaturase (FAD). Conventional plant breeding based high oleic acid peanut varieties through natural mutations on both genes are already commercialized (Yuan et al. 2019).

Soybean: Calyno oil represents one of the first GE crop product commercialized in USA. Compared to oil from conventional soybeans, Calyno oil contains about 80% oleic acid and up to 20% less saturated fatty acids. Amin et al. (2019), designed a single guided RNA sequence was designed with the help of several bioinformatics tools aimed to target distinct sites of *FAD2–2* omega 6 destuarase gene in soybean. A considerable enhancement in oleic acid content (65.58%) was observed, whereas the least level (16.08%) of linoleic acid were recorded.

Rice: Rice and other carbohydrate-heavy crops contain multiple kinds of starches. Some starches are readily converted into simple sugar and others like cellulose and xanthans are resistant to the digestive enzymes of human body and pass through without being digested and serve as a dietary fiber. High amylase with significant resistant starch content can be served as good dietary fiber in rice. CRISPR-mediated gene editing has yielded high-amylose rice, by editing Starch branching enzyme coding genes (*SBEIIb* and *SBEI*), which is a healthier option for overweight or diabetic individuals. The *SbeII* mutants showed significant increase in Amylose content and Resistant starch to 25% and 9.8%, respectively (Sun et al. 2017). Furthermore, disruption of *OsNramp5* gene led to the generation of low cadmium (Cd) rice lines. It is a key gene, which involved in absorption of trace elements such as Cd, Manganese (Mn), and other metal ions by root cells. The knockout of this gene significantly decreases the accumulation of Cd in rice grains from 2.90 to 0.05 mg/Kg (Tang et al. 2017).

Maize: Corteva Agriscience has come out with the first commercial GE crop i.e. new generation waxy maize developed by using CRISPR-Cas technology. Amylopectin starch is widely used in the food processing industry and production of adhesive and it is comparatively easy to process. The starch from commonly used maize kernels contains 25% amylose and 75% amylopectin, while that of waxy maize contains almost exclusively amylopectin (97%). Using CRISPR tools, the waxy gene (wx) was deleted in elite varieties, without affecting yield performance. The CRISPR technology offered the recapitulation of a *Wx1* whole gene deletion in one dozen elite inbreds with higher precision compared to the conventional approach (Friedrichs et al. 2019). These maize varieties are expected to be in the American market in a few years, pending field trials and regulatory testing.

10.3 Plant Defence-Based Traits

The principal agricultural challenges worldwide are biotic stresses, such as pathogens and pests. They cause 20–40% losses to global agricultural productivity thus posing significant challenges to food safety and security (Rodriguez-Moreno et al. 2017).

Rice: Rice Blast is considered as the most devastating disease in Rice and caused by *Magnaportheoryzae*. Stable non-race specific resistance to this disease has been shown by CRISPR targeted mutation in *OsERF922*, an ethylene responsive factor (Wang et al. 2016). The SWEET gene family of sugar transporters has been shown to play a vital role in multiple plant growth and developmental processes, including seed nutrition. The Rice bacterial blight disease-causing pathogen *Xanthomonas oryzae pv. Oryzae* infect the rice plant depends on expression of host (rice) gene, *OsSWEET13* (Zhou et al. 2015). OsSWEET transporters are responsible for host recognition and subsequent sugar acquisition by the bacterial pathogen *Xanthomonas oryzaepv. oryzae* during the rice bacterial blight disease. CRISPR-Cas9 mediated modification of SWEET11, SWEET13, and SWEET14 promoter confers broad-spectrum bacterial blight disease resistance in Indica rice, IR64 (Oliva et al. 2019).

Wheat: Several fungal diseases such as rust, smut, and powdery mildew cause severe yield losses and breeding efforts on to counter these diseases. Using genome editing technology, powdery mildew resistance in wheat is successfully developed. CRISPR/Cas9 mediated genetic alteration of three *EDR-1 genes, TaEDR1-1A, TaEDR1-1B*, and *TaEDR1-1D* shown high resistance to *Blumeriagraminis* f. sp. tritici infection (Zhang et al. 2017). Similar efforts were made in the past to generate knock out mutant of *TaMLO1* genes by using TALENS (Wang et al. 2014).

Cotton: Verticillium wilt is one the devastating diseases of cotton. The causative agent *Verticillium dahliae* Kleb, upon infection, causes leaf chlorosis, necrosis or wilting, leaf or boll abscission that results into significant loss in cotton yield (80%) and deterioration of fib quality (Wei et al. 2015). Recently Zhang et al. (2018a) has developed Verticillium disease resistance using gene editing approach, through targeted mutagenesis of*Gh14-3-3d* gene. GE plants showed enhanced resistance to verticillium wilt. It can be used as a resistant donor germplasm to breed disease-resistant cotton varieties and Hybrid.

10.4 Yield Improvement Traits

Wheat: The key wheat gene *TaGASR7-A1*has a positive impact on grain length under various growing environments (Lingli et al. 2014). Another important gene TaGW2 was identified by Zhang et al. (2018b) the effects on thousand-grain weight (TWG) and the mutations in the homoeologous gene led to changes in the grain length, width along with protein content. Mutations in both the yield genes were developed in two wheat variety by using CRISPR/Cas9 Ribonucleoprotein (RNP) complex system of gene editing. The off-target effect of CRISPR/Cas9 RNP complex was not observed.

Maize: Corteva Agriscience is making efforts to develop drought-tolerant maize using gene editing tools. This is being done with Jennifer Doudna's company, Caribou Biosciences 26. CRISPR-Cas was used to change promoter of the maize gene *ARGOS8* in such a way that it is transcribed more often, resulting in more *ARGOS8* protein in the cells (Shi et al. 2017), which is involved in the regulation of the plant stress hormone ethylene. Previous studies have shown that higher production of *ARGOS8* protein leads to a better yield under stressful growth conditions such as drought. The first field trials with the resulting maize hybrids have shown an increase in yield under drought stress compared to control plants, while yield performance was not affected by normal conditions.

Hybrid maize is a classical example of genetic improvement, which helped farmers from shifting from Doublecross to single hybrid coupled with high yields. The current Maize hybrid seed production system needs de-tasseling to avoid self-fertilization, as the female parents are yield fertile pollen. Thermosensitive genic male sterile system (TGMS) in maize is governed by *ZmTMS5* gene results in male sterility at high temperature. The *ZmTMS5* has been targeted for gene editing through CRISPR/Cas9 to create a desired allele for TGMS trait (Li et al. 2017).

Soybean: High yield is correlated to increase in number of "days to flowering" in short day (SD) plant-like soybean (Antonio et al. 1999). Induction of flowering switches depends on SD to long-day (LD) photoperiods. The expression of flowering hormone (florigen) governed by *FLOWERING LOCUS (FT)* in most of Angiosperm (Sun et al. 2011). CRISPR/CAS9 mediated genetic alterations in Soybean flowering time gene, GmFT2 yielded homozygous mutants, lacked flower reversion, and exhibited late flowering under both long-day and short-day conditions (Cai et al. 2018) (Table 10.1).

Rice:**Yield genes** Grain number: *Gn1a (Os01g0197700)*, Dense panicle: *DEP1 (Os09g0441900)*, Grain size: *GS3 (Os03g0407400)*, and Ideal Plant Architecture: *IPA1 (Os08g0509600)* genes have been edited using CRISPR-Cas9 technology. The homozygous mutants, *gn1a, dep1*, and *gs3* relates to enhanced grain number, dense erect panicles, and larger grain size, respectively. The *dep1* and *gs3* mutants exhibited novel phenotypes like semi-dwarf and grain with long awn. The *ipa1* mutants modified for the *OsmiR156* target region, showed two different phenotypes, fewer tillers or more tiller, implying that the resultant phenotype depends on the mutation in the targeted region (Li et al. 2016). This successful gene editing of these yield genes suggest that complex traits can be dissected out and desired mutations can be brought out in single variety.

Table 10.1 CRISPR directed field crop improvement

Crops	Nutrition	Yield	Defence traits
Rice	*SBEIIb* and *SBEI* (Sun et al. 2017), *OsNramp5* (Tang et al. 2017)	*Gn1a*, *DEP1, IPA1* (Li et al. 2016)	*OsSWEET11* (Jiang et al. 2013)
Corn	*Wx1* (Friedrichs et al. 2019)	*ARGOS8* (Shi et al. 2017) *ZmTMS5* (Li et al. 2017)	
Soybean	*FAD2* (Amin et al. 2019)	*GmFT2* (Cai et al. 2018)	*Avr4/6* (Fang and Tyler 2016)
Groundnut	*Ara h 1* to *Ara h 13* (Brazelton 2015) ahFAD2A and ahFAD2B (Yuan et al. 2019)		
Wheat	*α-gliadins* and *γ-gliadins* (Jouanin et al. 2019)	*TaGW2* and *TaGASR7* (Liang et al. 2018)	*TaEDR1—1A*, *TaEDR1—1B* and *TaEDR1—1D* (Zhang et al. 2017)
Cotton			*Gh14-3-3d* (Zhang et al. 2018a)

10.5 Conclusions

CRISPR mediated genome editing is a very powerful new breeding technique. This can create very precise and targeted mutations in the genome of the crop plants. Allelic series for a particular trait can be developed and can use in variety of improvement programs. The genome sequences and genetic annotations of most of the field crop plants are available on the public domains. Using this knowledge, CRISPR mediated genome editing can be used to develop desired traits in popular varieties and hybrids. Time taken for breeding of complex traits such as yield and drought can be reduced and make simple. CRISPR technology is proving very useful in improvement of crop yield with desired nutrition parameters in varieties and hybrids. Breeding for the disease resistance is another important task for plant breeders, where introgression of resistance genes from the wild sources to cultivated ones is a tedious job. CRISPR can help breeders to develop disease resistance traits in popular varieties and hybrids, which can avoid donor genetic content or genetic drag.

Websites

1. Using CRISPR to design superior foods, Jenna Gallegos, May 1, 2018 https://allianceforscience.cornell.edu/blog/2018/05/using-crispr-design-superior-foods/.
2. Statistics times 2018. https://statisticstimes.com/economy/countries-by-gdp-sector-composition.php.

3. Population growth.
4. https://www.nationalgeographic.com/foodfeatures/green-revolution/.
5. Enabling broad use of genome modification technologies to solve real world problems: A specific application in peanut. Thesis. Vincent A. Brazelton Jr. 2015. Iowa State University https://lib.dr.iastate.edu/cgi/viewcontent.cgi?article=5802&context=etd.

References

Amin N, Ahmad N, Wu N (2019) CRISPR-Cas9 mediated targeted disruption of *FAD2–2* microsomal omega-6 desaturase in soybean (Glycine max.L). BMC Biotechnol 19:9. https://doi.org/10.1186/s12896-019-0501-2

Antonio Carlos Baião de Oliveira, Carlos Sigueyuki Sediyama, Cosme Damião Cruz (1999) Selection for later flowering in soybean (Glycine max L.Merrill) F2 populations cultivated under short day conditions. Genetics Molecular Biology 22 (2):243–247

Brazelton, Vincent Antonio Jr., "Enabling broad use of genome modification technologies to solve real world problems: a specific application in peanut" (2015). Graduate Theses and Dissertations. 14795. https://lib.dr.iastate.edu/etd/14795

Cai Y, Chen L†, Liu X, Guo C, Sun S, Wu C, Jiang B, Han T, Hou W (2018) CRISPR/Cas9—mediated targeted mutagenesis of *GmFT2a* delays flowering time in soya bean. Plant Biotechnol J 16:176–185

Eshed Y, Lippman ZB (2019) Revolutions in agriculture chart a course for targeted breeding of old and new crops. Science eaax0025. https://doi.org/10.1126/science.aax0025

Fang Y & Tyler BM (2016) Efficient disruption and replacement of an effector gene in the oomycete Phytophthora sojae using CRISPR/Cas9. Mol Plant Pathol 17(1):127-39. https://doi.org/10.1111/mpp.12318

Friedrichs S, Takasu Y, Kearns P, Dagallier B, Oshima R, Schofield J, Moreddu C (2019) Meeting report of the OECD conference on "genome editing: applications in agriculture—implications for health, environment and regulation". Trans Res 28:419

Jiang W, Zhou H, Bi H, Fromm M, Yang B, Weeks DP (2013) Demonstration of CRISPR/sgRNA-mediated targeted gene modification in Arabidopsis, tobacco, sorghum and rice. Nucleic Acids Res. 41:e188 https://doi.org/10.1093/nar/gkt780

Jouanin A, Schaart JG, Boyd LA, Cockram J, Leigh FJ, Bates R, Wallington EJ, Richard GF. Visser RG, Smulders MJ (2019) Outlook for coeliac disease patients: towards bread wheat with hypoimmunogenic gluten by gene editing of α- and γ-gliadin gene families. BMC Plant Biol 19:333

Li M, Li X, Zhou Z, Wu P, Fang M, Pan X, Lin Q, Luo W, Wu G, Li H (2016) Reassessment of the four yield-related genes *Gn1a, DEP1, GS3,* and *IPA1* in Rice Using a CRISPR/Cas9 System. Front Plant Sci 7:377

Li J, Zhang H, Si X, Goa C (2017) Generation of thermosensitive male-sterile maize by targeted knockout of the *ZmTMS5* gene. J Genet Geno 44(9):465–468

Liang Z, Chen K, Zhang Y, Liu J, Yin K, Qiu JL et al (2018) Genome editing of bread wheat using biolistic delivery of CRISPR/Cas9 in vitro transcripts or ribonucleoproteins. Nat Protoc 13:413–430

Lingli D, Fangming W, Tao L, Daowen W et al (2014) Natural variation of *TaGASR7-A1* affects grain length in common wheat under multiple cultivation conditions. Mol Breed 34:937–947

Oliva R, Ji C, Atienza-Grande G, Huguet-Tapia JC, Perez-Quintero A, Li T, Eom J, Li C, Nguyen H, Liu B, Auguy F, Sciallano C, Luu VT, Dossa GS, Cunnac S, Schmidt SM, Slamet-Loedin

IH, Cruz CV, Szurek B, Frommer WB, White FF, Yang B (2019) Broad-spectrum resistance to bacterial blight in rice using genome editing. Nat Biotechnol 37:1344–1350

Ricciardi V, Ramankutty N, Mehrabia Z, Jarvisa L, Chookolingo B (2018) How much of the world's food do smallholders produce? Global Food Secur 7:64–72

Rodriguez-Moreno L, Song Y, Thomma BP (2017) Transfer and engineering of immune receptors to improve recognition capacities in crops. Curr Opin Plant Biol 38:42–49. https://doi.org/10.1016/j.pbi.2017.04.010

Shi J, Gao H, Wang H, Lafitte HR, Archibald RL, Yang M, Hakimi SM, Mo H, Habben JE (2017) *ARGOS8* variants generated by CRISPR-Cas9 improve maize grain yield under field drought stress conditions. Plant Biotechnol J 15:207–216. https://doi.org/10.1111/pbi.12603

Spielmeyer W, Ellis MH, Chandler PM (2002) Semidwarf (sd-1), "green revolution" rice, contains a defective gibberellin 20-oxidase gene. Proc Natl Acad Sci 99:9043–9048. https://doi.org/10.1073/pnas.132266399

Sun H, Jia Z, Cao D, Jiang B et al (2011) *GmFT2a*, a soybean homolog of Flowering Locus T, is involved in flowering transition and maintenance. PLoS ONE 6:12

Sun Y, Jiao G, Liu Z, Zhang X, Li J, Guo X, et al (2017). Generation of high-amylose rice through CRISPR/Cas9-mediated targeted mutagenesis of starch branching enzymes. Front. Plant Sci 8:298. https://doi.org/10.3389/fpls.2017.00298

Tang T, Mao B, Li Y, Lv Q, Zhang L, Chen C, He H, Wang W, Zeng X, Shao Y, Pan Y, Hu Y, Peng Y, Fu X, Li H, Xia S, Zhao B (2017) Knockout of OsNramp5 using the CRISPR/Cas9 system produces low Cd-accumulating indica rice without compromising yield. Scienti Rep 7:14438. https://doi.org/10.1038/s41598-017-14832-9

Wang Y, Cheng X, Shan Q, Zhang Y, Liu J, Gao C, Qiu JL (2014) Simultaneous editing of three homoeoalleles in hexaploid bread wheat confers heritable resistance to powdery mildew. Nat Biotechnol 32:947–951

Wang F, Wang C, Liu P, Lei C, Hao W, Gao Y, Liu Y, Zhao K (2016) Enhanced rice blast resistance by CRISPR/Cas9-targeted mutagenesis of the ERF transcription factor gene *OsERF922*. PLoS ONE 11 (4). https://doi.org/10.1371/journal.pone.0154027

Wei F, Fan R, Dong H, Shang W, Xu X, Zhu H, Yang J, Hu X (2015) Threshold microsclerotial inoculum for cotton Verticillium wilt through wet sieving and real time PCR. Phytopathology 105:220–229

Yuan M, Zhu J, Gong L, He L, Lee C, Han S, Chen C, Guohao, He G (2019) Mutagenesis of *FAD2* genes in peanut with CRISPR/Cas9 based gene editing BMC biotechnology 19:24. https://doi.org/10.1186/s12896-019-0516-8

Zhang Y, Bai Y, Wu G, Zou S, Chen Y, Gao C, Tang D (2017) Simultaneous modification of three homoeologs of *TaEDR1* by genome editing enhances powdery mildew resistance in wheat. Plant J 91:714–724

Zhang Z, Ge X, Luo X, Wang P, Fan Q, Hu G et al (2018a) Simultaneous editing of two copies ofGh14-3-3d confers enhanced transgene-clean plant defense against *Verticillium dahlia* in allotetraploid upland cotton. Front Plant Sci 842. https://doi.org/10.3389/fpls.2018.00842

Zhang Y, Li D, Zhang D, Zhao X, Cao X et al (2018b) Analysis of the functions of *TaGW2* homoeologs in wheat grain weight and protein content traits. Plant J 94:857–866

Zhou J, Peng Z, Long J, Sosso D, Liu B, Eom JS, Huang S, Liu S, Vera Cruz C, Frommer WB et al (2015) Gene targeting by the TAL effector PthXo2 reveals cryptic resistance gene for bacterial blight of rice. Plant J 82:632–643

Chapter 11
Patent Landscape of CRISPR/Cas

Paramita Ghosh

Abstract CRISPR/Cas technology is a widely used tool today and free to use for academic not-for-profit institutions. However, when it comes to developing commercial products using CRISPR/Cas technology, one has to obtain a license, as many over-lapping patents exist in this sphere. Single window licensing system could resolve such issues so that the use of CRISPR/Cas technology in developing new products is facilitated. Here, in this chapter, we look at the history, potential IP (intellectual property) players, patent scenario and describe the patent geographies and conflicts.

Keywords CRISPR/Cas · Patents · IP · USPTO · EPO · AddGene

11.1 Introduction

CRISPR/Cas is a molecular technology that has gained popularity in recent years and at the same time, has been surrounded by patent controversies. It is the simple yet robust technique for genome editing. CRISPR/Cas technology permit scientists to easily edit the DNA sequences and hence alter genetic code. It has the potential applications in correcting defects at the genetic level, treating disease and applied application in areas of crop improvement programmes. Even though this revolutionizing technology has gained its popularity in the recent decade, the discovery of CRISPR/Cas in prokaryotes dates back to the 1980s (Ishino et al. 1987).

P. Ghosh (✉)
Mahyco Research Centre, Aurangabad-Jalna Road, Dawalwadi, Jalna 431203, Maharashtra, India
e-mail: paramita.ghosh@mahyco.com

A. Bhattacharya et al. (eds.), *CRISPR/Cas Genome Editing*, Concepts and Strategies in Plant Sciences, https://doi.org/10.1007/978-3-030-42022-2_11

11.2 Discovery of GE System and Major Historical Players

GE research is sizable and involves scientists all over the globe. To understand the patentscape, we need to look into the series of discoveries which lead to applied application of CRISPR/Cas, from the humble beginning to understand the natural defence mechanism in bacteria to coming into researchers' lab as a potential gene editing tool and then becoming the most popular editing technology of today (Gasiunas et al 2012). CRISPR has gone through a lot of controversies, be it related to patent or regulations along with an increasing diversification of related Cas systems. CRISPR/Cas was first discovered in 1987 in Osaka University in Japan by Yoshizumi when it was accidentally cloned along with the "iap" gene (Ishino et al. 1987). Independently, researchers have participated in the discovery of the aspects of congregate of sequence repeats. Researchers from Netherland published two articles about a group of fragmented undeviating duplicates in the organism *Mycobacterium tuberculosis* (Soolingen et al. 1993). At the same time, Francisco Mojica, a researcher in Spain studied the function of the group of repeats in the *Achaeal* organisms of Haloferax and Haloarcula species, and their function was studied (see, Mojica et al. 2000). The significance of such repeats remained unknown till 2007.

Emmanuelle Charpentier while working for Uni of Umea made a significant finding about the function of the Cas9 enzyme (see Deltcheva et al. 2011). She along with Martin Jinek and Jennifer Daudna, of UC, Berkeley (UCB) cracked the puzzle and showed for the first time, the significance in genome editing revolutionizing the field (Jinek et al. 2012). The finding were reported in two publications in the Journal of Science and PNAS in the year 2012. They showed how GE from bacteria can be transformed into a simple gene editing system (Jinek et al. 2012). While Feng Zhang and team at the Board Institute in Boston and MIT advanced the CRISPR research in eukaryotes (Cong et al. 2013). At the same time, George Church and co-workers at Harvard exhibited the use in human cells (Byrne et al. 2014).

11.3 The CRISPR Patent Dispute

The last four decennia are the witness of not only the advancement in biotechnology research but also a sharp rise in intellectual property protection like patent filling where Universities and public research organizations made vast contribution in making such applications. The 1980 Bayh-Dole Act allows US universities to hold IP rights (form more info please refer, website 1). UCB had filled its first CRISPR patent in May 2012 (US Patent Application No. 61/652,086), was based on ex vivo work including methods and systems for alter given sequence without restriction to the use in higher organisms. Broad Institute, had also filed a patent in the same year 2012, only after UCB had filed their patent. Broad Institute opted for an expedited review and fast tacked the examination that got its application granted first (US Patent Application No. 61/842,322). The Broad patent for genome editing and uses in complex cells including that from plants, animals, humans, was given in April 2014 (US **8,697,359** *B1*). The patent fight began soon after the USPO granted the Broad Institute a patent for higher organisms. The University of California had indecision and arguments against the Broad's interference with that of own patent. Thus an interference proceeding started between Broad and UCB. To establish invention dates, USPTO were to collect, consider, and compare historical documentary evidence. The patent rights were to be accorded under the early "first to invent" order rather than "first to file" patent system as both the applications were made before 2013 (Website 2). The argument of the University of California team was based on the "obviousness" of the invention in eukaryotes even though the patent UC applied for overtly described application in prokaryotes.

The CRISPR battle took a turn when the USPO supported Broad in early, 2017. Consequently, the UCB immediately appealed to the US Court of Appeals for the Federal Circuit, professing as a "fundamental errors of law" that would give Broad significant advantage. Continuing the fight, The Broad had filed its own brief in October in response to UC's appeal which claimed that there was lack of "substantial evidence" to support its findings that the PTAB had made a material legal error that the PTAB failed to consider certain evidence when it closed the interference case.

The United States Court of Appeals for the Federal Circuit on 10th September 2018 conceded a judgement from PTAB which awarded crucial intellectual property rights to Broad (Website 2, 3). The EPO Board of Appeal's recently repealed GE patent, EP2771468 and others of the Broad.

11.4 A Broad Landscape

The CRISRP patent estate has many players (Sherkow 2018). The earliest U.S patent applications related to or which introduce CRISPR in any manner go before to 2004 (US 7919277), almost a decade before the controversy actually begun. One of the patent (U.S. patent No. 7919277) belonging to The Danish food chemistry company, Danisco A/S where they had prerogative of methods to characterize certain sectors in the GE-CRISPR regions of bacteria used in yogurt production (US 7919277). There were few more patents related to CRISPR that followed in the same era, were filed by Barrangou and his collaborators which had restricted the claims to typing of bacterial strain, resistance to phage and utilization in bacteria (US 8361725, US 9399801 and US 9816140). Another important CRISPR-patent relating to guide sequence is to inactivate a target polynucleotide was by Rebecca Terms, Micheael Terms and Caryn Hal from University of Georgia Research foundation Inc., where they had confined their studies to only prokaryotic microbes in vitro (US 8546553). The only striking similarity is CRISPR-Cas9 component and gRNAs. Feng Zhang along with his colleagues at the Board Institute has filed a number of innovations in the mechanism to deliver the GE cargo whereas Doudna and Charpentier have none. Therefore, US8697359 patent is a combination of multiple technology categories including enzyme component gRNA and Cas9, cleavage activity of Cas, liposome and nano-particle delivery and also in therapeutic applications.

A consulting firm based in Switzerland which specializes in IP Studies has gathered data that reflects that an enormous number of patent applications are filed/pending worldwide (excess of 12,000) which can be classified into 4,600 patent families (please refer to the-scientist.com/news-opinion/a-brief-guide-to-the-current-crispr-landscape—66,128, for more details). Apart from the Broad and UCB, the Chinese Academy of Sciences depicts many other patent families in GE. The other prominent companies perusing this field are DuPont, and Editas Medicine (co-founded by Broad and UC) (Website 4). The data shows that the rise in CRISPR patent steeped up after 2012. The top players being the leading academic institutions and corporations, most notably the Boston academic cluster. Meanwhile, UC has limited patent holdings (Egelie et al. 2016). In the field of agriculture, the top players are China Agriculture University, Huazhong Agriculture University, Institute of Crop Sciences, Chinese Academy of Agriculture, South China Agriculture University, Institute of Genetic and Developmental Biology, and Dow Agrosciences (Corteva). In terms of the number of patents filed, China tops the list.

The Figs. 11.1, 11.2, 11.3 and 11.4 were generated through data analysis done in Orbit database software.

An analysis of patent filed in the field of agriculture was done using orbit database using simple keywords CRISPR and agriculture showed over one thousand patents filed in the field of agriculture. The trend shows a steep rise after 2012. The analysis highlights, compared to just two patents filed in the year 2012; 70 patent applications were filed for new inventions in the field of agriculture in 2014 and, 169 patent applications in the year 2016. The top players are China Agriculture University, Huazhong Agriculture University, Institute of Crop Sciences, Chinese Academy of Agriculture, South China Agriculture University, Institute of Genetic and Developmental Biology and Dow Agrosciences (Corteva). In terms of the number of patents filed, China tops the list.

A notable increase in patent filing is observed from the year 2012 to 2017. An 18 month lag period may have resulted in many applications being missed in the last year, thus showing a dip in the graph.

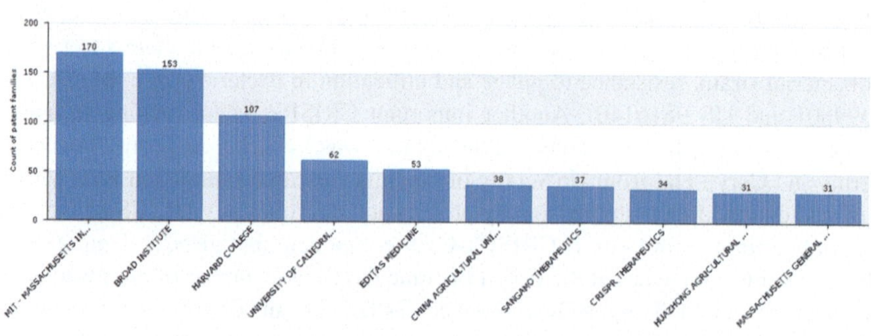

Fig. 11.1 The CRISPR patent landscape has expanded widely in recent years

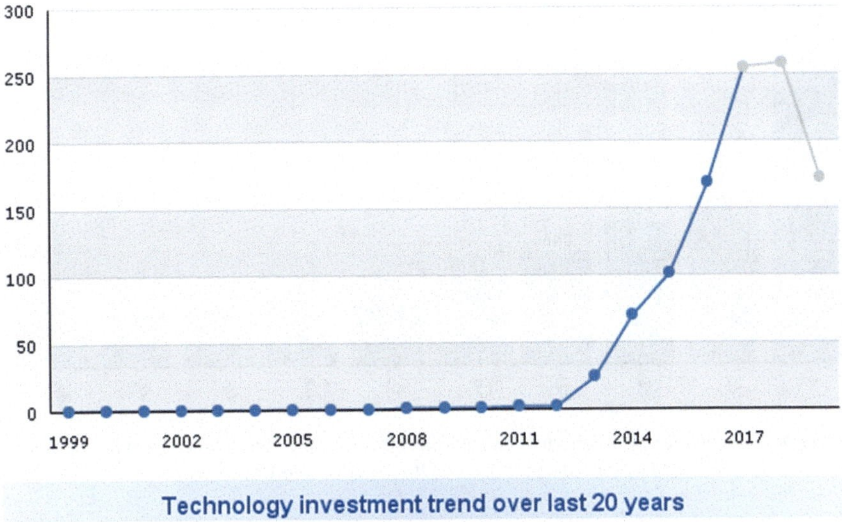

Fig. 11.2 The number of CRISPR patents in the field of agriculture, searched on the basis of "CRISPR and agriculture" in Orbit Innovations index

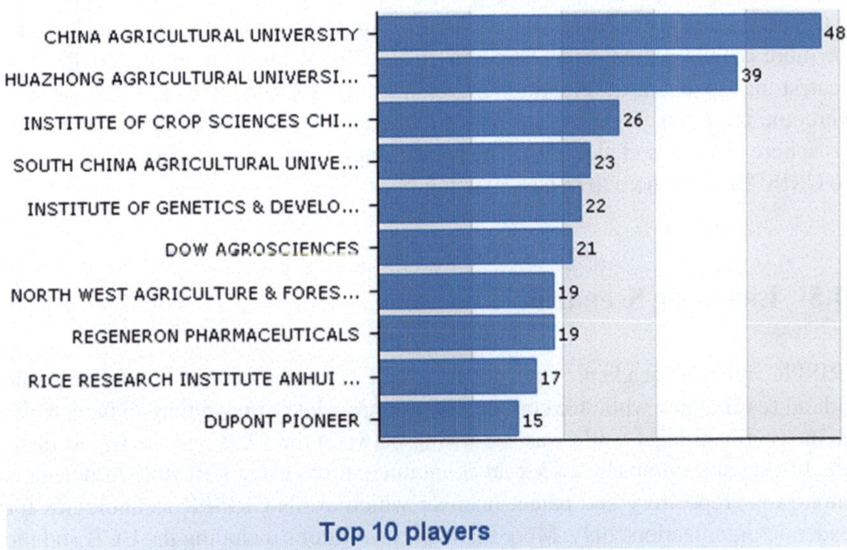

Fig. 11.3 Overview of the top 10 GE players

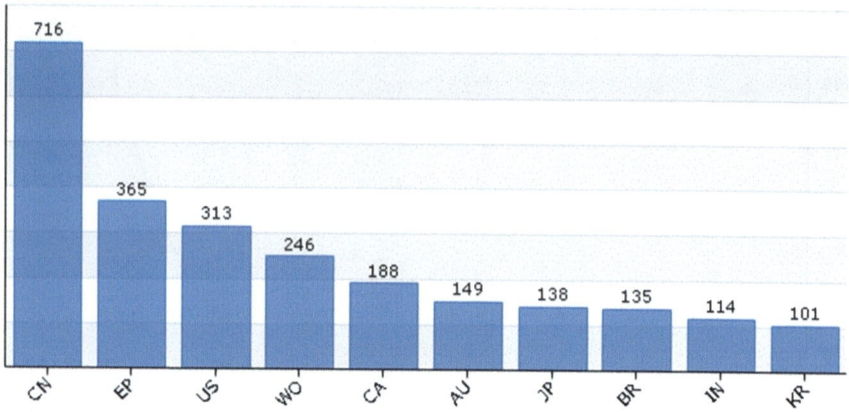

Fig. 11.4 Geographical distribution of patent filings in the field of agriculture. Most inventions are filed in China, followed by the European Patent Office (EPO) and United States

While undertaking search using the software, the keywords searched for were CRISPR and agriculture. Again, family patents are defined as single invention/family. This includes granted and proclaimed ones.

Over the years, the holdings of CRISPR enzymes have grown by leaps and bounds. For more details please refer, Burstein et al. (2017), Sherkow et al. (2018). The greatest incentives for developing new versions of GE enzymes would have been to overcome numerous legal hassles and the liberty to operate without restrictions in this sphere (Sherkow et al. 2018). The legal complexity is also driving innovation in the CRISPR/Cas sphere to bypass existing ones.

11.5 Licensing Scenario

CRISPR rights to academic and not-for-profit organizations are freely available without restrictions; while for commercial use has a lot of uncertainty to face. while the international legal battle was seesawing between the UCB and the Broad Institute, life science companies, seek an economic path to using CRISPR. AddGene is a non-profit repository and patent licensor which avails CRISPR technologies for academic organizations only. More than 700 institutions including the UCB and the Board Institute have arranged to make their CRISPR construct and intellectual property accessible through AddGene (Website 5). However, the challenge remains for commercial entities including start-ups who would like to bring products to market. After the kick start of the patent filing on CRISPR system after 2012, very few commercial assignees were in the field. Dow Dupont (now Corteva) is only the large corporate within the top ten, holding 33 inventions, in genome editing and dominating the major agricultural field of application after acquiring Danisco in 2015 (Knut et al. 2016). Recently, Broad Institute made a robust effort in the area of cross-licensing together with partners, please see, website 6. In fact, the Broad has made several licensing deals with Arcadia Biosciences to develop traits related to nutritional and agricultural productivity in its core crops, and with Syngenta for use in multiple crops (Website 7). The Broad Institute and Massachusetts General Hospital have also signed agreements with Pairwise, a agriculture and food company, to license CRISPR genome editing technologies to develop new applications in crop editing, in food research and to increase the sustainability of modern agriculture (website 8).

11.6 Conclusion

CRISPR/Cas as a genome editing system comes with high promises like the other biotechnology tools of the past and present. However, for its successful commercial adoption, regulatory scenarios and consumer acceptability will play a major role. Future of CRISPR will be decided ultimately by the stakeholders, administrators and researchers alike. Ease of doing business amidst of over-lapping patents will be critical. A single licensing window can solve many of the challenges individual organization face today while using CRISPR/Cas for developing new agricultural products or varieties.

References

Burstein D, Harrington LB, Strutt SC, Probst AJ, Anantharaman K, Thomas BC, Doudna JA, Banfield JF (2017) New CRISPR-Cas systems from uncultivated microbes. Nature 542:237–241

Byrne SM, Mali P, Church GM (2014) Genome editing in human stem cells. Methods Enzymol 546:119–138

Cong L, Ran FA, Cox D, Lin S, Barreto R, Habib N, Hsu PD, Wu X, Jiang W, Marraffini LA, Zhang F (2013) Multiplex genome engineering using CRISPR/Cas systems. Science 339:819–823

Deltcheva E, Chylinski K, Sharma CM, Gonzales K, Chao Y, Pirzada ZA, Eckert MR, Vogel J, Charpentier E (2011) CRISPR RNA maturation by *trans*-encoded small RNA and host factor RNase III. Nature 471:602–607

Egelie KJ, Graff GD, Strand SP, Johansen B (2016) The emerging patent landscape of CRISPR-Cas gene editing technology. Nat Biotechnol 34:1025–1031

Gasiunas G, Barrangou R, Horvath P, Siksnys V (2012) Cas9-crRNA ribonucleoprotein complex mediates specific DNA cleavage for adaptive immunity in bacteria. PNAS 109:2579–2586

Ishino Y, Shinagawa H, Makino K, Amemura M, Nakata S (1987) Nucleotide sequence of the iap gene, responsible for alkaline phosphatase isozyme conversion in Escherichia coli, and identification of the gene product. J Bacteriol 169:5429–5433

Jinek M, Chylinski K, Fonfara I, Hauer M, Doudna JA, Charpentier E (2012) A programmable dual-RNA–guided DNA endonuclease in adaptive bacterial immunity. Science 337:816–821

Mojica FJ, Diez-Villasenor C, Soria E, Juez G (2000) Biological significance of a family of regularly spaced repeats in the genomes of *Archaea*, Bacteria and Mitochondria. Mole Microbiol 36:244–246

Sherkow JS (2018) The CRISPR patent landscape: past, present and future. CRISPR J 1:5–9

Van Soolingen D, de Haas PE, Hermans PW, Groenen PM, van Embden JD (1993) Comparison of various repetitive DNA elements as genetic markers for strain differentiation and epidemiology of *Mycobacterium tuberculosis*. J Clin Microbiol 31:1987–1995

Websites

Website 1: U.S. Code Chapter 18—Patent Rights in Inventions Made with Federal Assistance. cornell.edu

Website 2: Bitter CRISPR patent war intensifies Heidi Ledford Nature News 26 October 2017 (Nature https://doi.org/10.1038/nature.2017.22892)

Website 3: Broad Institute takes a hit in European CRISPR patent struggle; Kelly Servick. Jan. 18, 2018, 3:30 PM

Website 4: A Brief Guide to the Current CRISPR Landscape; Jul 15, 2019

Website 5: Depositing Institutions, AddGene, https://www.addgene.org/browse/institutions/ (last accessed November 18, 2019)

Website 6: Cameron D. DuPont Pioneer and Broad Institute Join Forces to Enable Democratic CRISPR Licensing in Agriculture. https://www.broadinstitute.org/news/dupont-pioneer-and-broad-institute-join-forces-enable-democratic-crispr-licensing-agriculture

Website 7: https://www.genomeweb.com/business-news/federal-court-hands-broad-institute-victory-crispr-patent-fight-against-uc-berkeley#.Xfc4LNSF7Dc. Federal Court Hands Broad Institute Victory in CRISPR Patent Fight Against UC Berkeley

Website 8: Broad Institute, Massachusetts general hospital license CRISPR tech to pairwise. Mar 18, 2019

Patents

US Patent Application No. 61/652,086 (US20140068797)
US Patent Application No. 61/842,322 (Us *8,697,359*)
US 7919277
US 8361725
US 9399801
US 9816140
US 8546553

The manufacturer's authorised representative in the EU is Springer Nature Customer Service Centre GmbH, Europaplatz 3, 69115 Heidelberg, Germany. If you have any concerns regarding our products, please contact ProductSafety@springernature.com

Printed and bound by CPI Group (UK) Ltd, Croydon, CR0 4YY
22/09/2026
02221063-0013